Nanoscale Self-Assembly: Nanopatterning and Metrology

Nanoscale Self-Assembly:
Nanopatterning and Metrology

Editor

Federico Ferrarese Lupi

MDPI • Basel • Beijing • Wuhan • Barcelona • Belgrade • Manchester • Tokyo • Cluj • Tianjin

Editor
Federico Ferrarese Lupi
Advanced Materials &
Life Sciences
Istituto Nazionale di
Ricerca Metrologica
Torino
Italy

Editorial Office
MDPI
St. Alban-Anlage 66
4052 Basel, Switzerland

This is a reprint of articles from the Special Issue published online in the open access journal *Nanomaterials* (ISSN 2079-4991) (available at: www.mdpi.com/journal/nanomaterials/special_issues/Self-Assembly_Nanopatterning).

For citation purposes, cite each article independently as indicated on the article page online and as indicated below:

LastName, A.A.; LastName, B.B.; LastName, C.C. Article Title. *Journal Name* **Year**, *Volume Number*, Page Range.

ISBN 978-3-0365-1961-6 (Hbk)
ISBN 978-3-0365-1960-9 (PDF)

Contents

About the Editor

Federico Ferrarese Lupi

Federico Ferrarese Lupi graduated in Physics of Advanced Technologies at the department of physics of the University of Turin in 2008. In 2012, he received his PhD in Physics "Cum Laude" at the University of Barcelona. Between 2012 and 2015, he worked as postdoctoral fellow at the IMM–CNR institute. During that period, he focused on the fabrication of self-assembled nano-structures using block copolymers and its integration with the next generation of electronic devices. Currently, he is a researcher at the Italian National Metrology Institute. He is involved in several projects covering different aspects of the fabrication and characterization of materials at the nanoscale. His main research activity is focused on the optical characterization and nano-patterning of polymeric and silicon-based materials with potential application in nanometrology. Federico Ferrarese Lupi is author or co-author of more than 60 scientific papers and 2 patents.

Preface to "Nanoscale Self-Assembly: Nanopatterning and Metrology"

The self-assembly process underlies a plethora of natural phenomena from the macro to the nano scale. Often, technological development has found great inspiration in the natural world, as evidenced by numerous fabrication techniques based on self-assembly (SA). One striking example is given by epitaxial growths, in which atoms represent the building blocks. In lithography, the use of self-assembling materials is considered an extremely promising patterning option to overcome the size scale limitations imposed by the conventional photolithographic methods. To this purpose, in the last two decades several supramolecular self-assembling materials have been investigated and successfully applied to create patterns at a nanometric scale (e.g., spherical colloids, block copolymers, nanotubes, and nanowires).

Although considerable progress has been made so far in the control of self-assembly processes applied to nanolithography, a number of unresolved problems related to the reproducibility and metrology of the self-assembled features are still open. Addressing these issues is mandatory in order to allow the widespread diffusion of self-assembling materials for applications such as microelectronics, photonics, or biology.

In this context, the aim of the present Special Issue is to gather original research papers and comprehensive reviews covering various aspects of the self-assembly processes applied to nanopatterning. Topics include the development of novel SA methods, the realization of nanometric structures and devices, and the improvement of their long-range order. Moreover, metrology issues related to the nanoscale characterization of self-assembled structures are addressed.

The directed self-assembly (DSA) is a process whereby the long-range ordering of a self-assembling system is aided by the presence of chemical or physical directing agents, such as periodic templates, surface functionalizations or external fields. In the first chapter, Tommaso Jacopo Giammaria, Ahmed Gharbi, Anne Paquet, Paul Nealey and Raluca Tiron report a novel resist-free chemo-epitaxy process allowing the DSA of symmetric block copolymers (BCPs). The chemical pattern that permits the DSA is composed of polystyrene guides prepared by 193i lithography and a grafted PS-r-PMMA brush layer. The reported process is fully compatible with 300 mm clean-room facilities, and the performance in terms of roughness can be considered at the same level to the state-of-the-art methods reported in the literature.

The DSA process is not exclusively linked to BCPs, but can easily be extended to other self-assembling materials such as colloidal nanospheres. In this regard, Eleonora Cara and coworkers performed a systematic study on the effect of the spin-coating conditions on the SA process of colloidal nanospheres inside pre-patterned templates prepared by direct laser-writer lithography. This study revealed that in the best deposition conditions, the SA leads to a well-ordered nanospheres monolayer that extended for tens of micrometres.

The mandatory reduction of the defects of SA nanostructures, which prompted the development of the DSA techniques, is only one of the hurdles limiting the diffusion of SA materials for lithographic purposes. Another important problem of self-assembling polymeric materials is linked to their intrinsic fragility to the conventional etching processes, which limits the maximum aspect ratio obtainable in the pattern transfer process. The development of new synthetic approaches for large-scale SA materials with enhanced performances is therefore required in the pursuit of the fabrication of next-generation devices. SA materials as BCPs are ideal candidates for the selective incorporation of a variety of inorganic materials through the sequential infiltration synthesis (SIS) technique. In the comprehensive review reported in the third chapter, Cara and coworkers report the latest advances in nanostructured inorganic materials synthesized by infiltration of BCPs, and their emerging electrical and optical properties.

The close link between technological development and advancement in measurement techniques has been established for a very long time. This concept is best summed up in the sentence "you cannot measure it, you cannot improve it", expressed by Lord Kelvin i in the late 1800s [1]. The understanding of self-assembly phenomena is also subject to the continuous improvement of metrology. In chapter four, Julius Bürger, Vinay S. Kunnathully, Daniel Kool, Jörg K. N. Lindner and Katharina Brassat use advanced analytical (scanning) transmission electron microscopy ((S)TEM), electron energy loss spectroscopy (EELS) and energy filtered TEM (EFTEM), to provide information on cylinder- and lamellae-forming BCPs morphology and interfaces at highest resolution. This methods allows correlating the internal structure of the SA nano-domains with line edge roughness, interface widths, and domain sizes.

In recent years the SA of BCPs in aqueous solution is gaining considerable importance in modern polymer science. In chapter 5, Noah Al Nakeeb, Ivo Nischang and Bernhard V.K.J. Schmidt investigate the self-assembly properties of double hydrophilic block copolymers (DHBC) in aqueous solution via dynamic light scattering (DLS) and cryogenic scanning electron microscopy (cryo-SEM). This combination of measurement techniques allowed them to report on the self-assembly process of completely hydrophilic structures in solution formed from fully biocompatible building entities in water.

In chapter 6, AFM and photoluminescence (PL) spectroscopy are resorted to in order demonstrate the correlation between deposition speed and photoluminescence quenching in polyfullerenes microstructures and donor polymer films fabricated by convective self-assembly. The precise control of deposition speed during the fabrication procedure leads to an optimized film microstructure comprised of interconnected crystalline polymer domains comparable to molecular dimensions intercalated with similar polyfullerene domains.

Differential centrifugal sedimentation (DCS) enables direct measurements of differential size distributions of complex nanoparticle mixtures. This technique is also suitable for resolving particle populations, which only slightly differ in size or density. Reliable data on nanoparticle agglomeration state are crucial for hazard and risk assessment of nanomaterials and for grouping and read-across of nanoforms. In the final chapter, Claudia Simone Plüisch, Rouven Stuckert and Alexander Wittemann determine the sedimentation coefficient distributions supracolloidal assemblies by DCS. Furthermore, the practical implementation of the analytical findings into preparative centrifugal separations is explored.

In the last decade several efforts have been focused on the realization of materials and devices with nanometric feature size. The vast majority of them are conceived in a simple planar configuration. However, the growing demand for more complex materials, useful for example for the realization of micro- and nano-scale multifuctional devices or metamaterials, have pushed the development of novel nanofabrication techniques, capable of removing challenges and limits of traditional self-assembly techniques. In chapter 8, Fengqiang Zhang and coworkers demonstrate a simple and effective 3D printing strategy to achieve the nano-resolution printing of multiple materials via layer-by-layer deposition. In less than two hours, large-scale concentric ring arrays with minimum feature size below 50 nm were printed, verifying the capacity of high-throughput, high-resolution, and rapidity of the proposed technique.

Reference

[1] William Thomson Baron Kelvin, "Popular Lectures and Addresses: Constitution of matter I" 1891.

Federico Ferrarese Lupi
Editor

 *nanomaterials*

Article

Resist-Free Directed Self-Assembly Chemo-Epitaxy Approach for Line/Space Patterning

Tommaso Jacopo Giammaria [1,*], Ahmed Gharbi [1], Anne Paquet [1], Paul Nealey [2] and Raluca Tiron [1]

[1] Université Grenoble Alpes, CEA, Leti, F-38000 Grenoble, France; ahmed.gharbi@cea.fr (A.G.); anne.paquet74@gmail.com (A.P.); raluca.tiron@cea.fr (R.T.)
[2] Institute for Molecular Engineering, University of Chicago, 5747 South Ellis Avenue, Chicago, IL 60637, USA; nealey@uchicago.edu
* Correspondence: tommaso.giammaria@cea.fr

Received: 15 November 2020; Accepted: 5 December 2020; Published: 7 December 2020

Abstract: This work reports a novel, simple, and resist-free chemo-epitaxy process permitting the directed self-assembly (DSA) of lamella polystyrene-block-polymethylmethacrylate (PS-*b*-PMMA) block copolymers (BCPs) on a 300 mm wafer. 193i lithography is used to manufacture topographical guiding silicon oxide line/space patterns. The critical dimension (CD) of the silicon oxide line obtained can be easily trimmed by means of wet or dry etching: it allows a good control of the CD that permits finely tuning the guideline and the background dimensions. The chemical pattern that permits the DSA of the BCP is formed by a polystyrene (PS) guide and brush layers obtained with the grafting of the neutral layer polystyrene-random-polymethylmethacrylate (PS-*r*-PMMA). Moreover, data regarding the line edge roughness (LER) and line width roughness (LWR) are discussed with reference to the literature and to the stringent requirements of semiconductor technology.

Keywords: directed self-assembly (DSA); block copolymers (BCPs); chemo-epitaxy; polystyrene-block-polymethylmethacrylate (PS-*b*-PMMA); line/space patterning; line edge roughness (LER); line width roughness (LWR)

---

1. Introduction

The block copolymers (BCPs) have attracted more and more attention because under suitable conditions, they self-assemble in highly ordered polymeric templates with well-defined sub-20 nm periodic features that could be extremely useful for a wide range of nanotechnology applications [1–4]. The morphology of the self-assembled nanostructures (spherical, cylindrical, gyroid, or lamellar) is determined by the fraction of one of the blocks with respect to the other one that form the BCP chain [5]. The directed self-assembly (DSA) of block copolymers (BCPs) permits obtaining well-ordered nanostructures having high resolution [6]. In this context, the chemo-epitaxial process represents one of the most used approaches to induce the long-range ordering of the BCPs nanostructures by means of the chemical contrast between the guideline and background. In this case, it is possible to obtain a long-range ordering when the dimension of the guideline and background are commensurate with the intrinsic period of the BCP [7–10]. Numerous methods were developed for the DSA of lamellae forming BCPs [11–14], but in the framework of 300 mm fab process manufacturing, where all the fabrication processes are made on a 300 mm wafer, LiNe [15], SMART [16], and COOL [17] processes are the most attractive.

The LiNe process is based on the trimming of chemical patterns obtained by argon fluoride (ArF) immersion lithography. After the deposition and cross-link of the PS layer, positive-tone photoresist patterns are created on top of it by ArF immersion lithography. At this point, the pattern consisting of the photoresist lines is trimmed, transferred into the cross-linked PS layer, and stripped with a strong solvent. Subsequently, a neutral layer is deposited and grafted on the substrate. In this way, the chemical patterns consisting of PS guidelines and a grafted neutral layer background are formed. Finally, the BCP thin film is deposited on the chemical pattern and thermally treated in order to achieve the phase separation in perpendicularly oriented lamellae [15].

Regarding the SMART process, a photo-resist layer is deposited on top of the PS-*r*-PMMA cross-linked layer. On top of this photo-resist layer, a lithography process is performed in order to transfer the pattern through the PS-*r*-PMMA layer by means of dry etching. After the stripping of the photoresist, a pattern consisting of zones with a PS-*r*-PMMA layer and zones where the substrate is exposed are formed. Subsequently, selective grafting of the functionalized polymer is performed in order to create the guides affine to one of the two blocks of the PS-*b*-PMMA BCP in order to achieve the DSA of the lamellae when the PS-*b*-PMMA layer is deposited on it [16].

Finally, for what concerns the COOL process, the guide pattern consists of line/space structures obtained by ArF immersion lithography of a photoresist. Subsequently, the resist pattern is etched in order to slim the guides and slightly etch the substrate. This etching step modifies the surface energy of the resist guide, making it more affine to the PMMA block of the BCP. In this way, the resist guide acts as a directional guide for the subsequent DSA of the lamellae BCP. At this point, the selective grafting of functionalized PS-*r*-PMMA is performed. The grafting process takes place exclusively on the substrate, leaving the properties of the guides unaltered. In the end, the BCP layer is deposited and baked on the top of the chemical pattern. In this case, the resist guides act as directional guidelines for the DSA [17].

The principal limitation of these methods in the integration of the DSA BCPs chemo-epitaxy process in 300 mm fab is related to the use of polymeric pining materials to fabricate the guideline for the DSA. This can limit the control of the surface free energy and the final critical dimension of the guidelines [17,18].

This work reports a novel, simple, and resist-free chemo-epitaxy process permitting the DSA of lamella forming BCPs on the 300 mm wafer. This process is called the "Trim-Ox approach". Here, conventional lithography is used to manufacture topographical tetraethyl orthosilicate (TEOS) line/space patterns. The TEOS lines were exploited to create the PS guidelines that permit achieving the DSA. The critical dimension (CD) of the obtained TEOS lines can be easily trimmed by means of wet etching: it allows a good control of the CD that permits finely tuning the guideline and the background dimensions. The evolution of key metrics—period (L), LER, LWR, and Orientational Parameter—are evaluated as a function of the TEOS pitch size. The possibility of obtaining a sub-20 nm critical dimension guideline allows the integration of BCPs having a period below 20 nm.

2. Experimental

2.1. Materials

The BCP studied was a lamella-forming poly(styrene-block-methylmethacrylate) (PS-*b*-PMMA) formulated in propylene glycol methyl ether acetate (PGMEA) that was synthesized by ARKEMA under the trend name Nanostrength® EO. The BCP has an intrinsic period of 32 nm [18], and it is referred to as L32 in this work. The guideline was a functionalized PS that can graft the substrate to form a compact brush layer. Four different polystyrene-random-methylmethacrylate (PS-*r*-PMMA) backgrounds were used:

- Cross-linked PS-*r*-PMMA (NLa) layer
- Functionalized PS-*r*-PMMA (NLb) with high molar mass
- Functionalized PS-*r*-PMMA (NLc) with low molar mass
- Functionalized PS-*r*-PMMA (NLd) with low molar mass (comparable to NLc) and different fractions of PS with respect to NLc.

All polymers were synthetized by the industrials partners, Arkema and Brewer Science, and used as received.

For the lithography stacks, silicon anti-reflective coatings (SiARC), spin-on-carbon (SOC), and tetraethyl orthosilicate (TEOS) layers were employed. Additionally, a lithography step was realized with a chemically amplified negative resist developed in the negative tone using TMAH.

All process steps presented thereafter were performed on the LETI's 300 mm pilot line. More precisely, coating, annealing, and wet treatment were carried out on the DSA-dedicated SCREEN RF3 and Sokudo Duo tracks.

The etching process with hydrofluoric (HF) acid 1% was performed by using the RAIDER 4B tool for 300 mm wafer.

2.2. Characterization

The top-view critical dimension scanning electron microscopy (CD-SEM) images were obtained using a HCG4000 CD-SEM from Hitachi with an accelerating voltage of 500 V. The fraction of the perpendicularly oriented lamellae and intrinsic period (L) of the BCP were obtained by means of the analysis of the CD-SEM images with ImageJ software. LER, LWR, and Herman's orientational parameter (P) were obtained by means of ADAblock software [19].

The film thicknesses measurement was performed by ellipsometry methods with an Atlas XP+ tool from Nanometrics.

3. Results

Figure 1 summarized the process steps for the DSA of the lamellar PS-*b*-PMMA on patterned 300 mm wafers. A stack consisting of tetraethyl orthosilicate (TEOS)/spin-on carbon (SOC)/silicon anti-reflective coating (SiARC)/lithography resist is deposited on 30 nm thick titanium nitride (TiN) substrate (Figure 1a). Then, a 140 nm TEOS line/space pattern (Figure 1b) was obtained by means of 193i lithography on the TiN substrate. The pitch of the TEOS lines varies from 97.5 ± 1 nm to 200 ± 1 nm with a step of 2.5 nm, while the CD of the lines can be easily optimized by means of HF etching at 1% (Figure 1c). At this point, the neutral layer PS-*r*-PMMA is deposited by means of the spin-coating method (Figure 1d), TEOS lines are removed by HF etching at 1% (Figure 1e), and PGMEA rinsing is performed. Subsequently, functionalized PS is selectively grafted to form the guideline and rinsed with PGMEA (Figure 1f). Finally, the BCP L32 is deposited by spin-coating and annealed in order to obtain the self-assembly (Figure 1h).

Figure 2 shows the lateral etching rate evolution of TEOS lines in line/space pattern having a pitch of 120 ± 1 nm measured from top-view CD-SEM images where the CD of the TEOS lines pass progressively from 82 ± 2 nm to 15 ± 1 nm as a function of the HF etching time. This process step is referred to the lines trimming sketched in Figure 1c. The final etching rate is 0.25 nm/s (15 nm/min) on 300 mm wafer, which is three times faster than standard silicon dioxide (SiO_2) [20].

The CD of the TEOS lines defines the CD of the PS guiding lines. In order to obtain a long-range ordering of the BCP L32, the CD of the lines must be optimized at 15 nm, which represents the natural half period ($L_0/2$) of one BCP lamella [9]. For this reason, the HF etching time was fixed at 140 s.

At this point, the 300 mm wafers are spin coated with the selected neutral layers by using the SCREEN RF3 track and then annealed by using the SCREEN Sokudo DUO track. Figure 3 shows a representative plan view CD-SEM images of steps d, e, f, and g sketched in Figure 1.

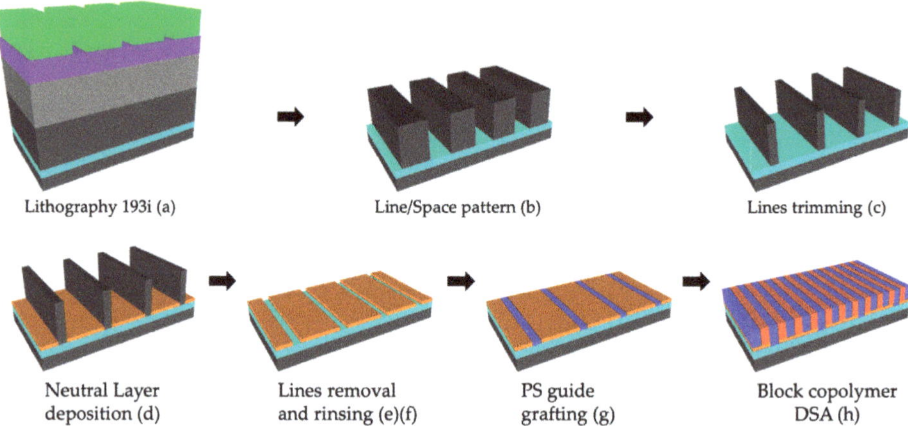

Figure 1. Schematic representation of trim-ox process flow. (**a**) Lithography step where the stack consists of, starting from the bottom: tetraethyl orthosilicate (TEOS) (140 nm)/titanium nitride (TiN) (30 nm)/TEOS (140 nm)/spin-on-carbon (SOC) (140 nm)/silicon anti-reflective coatings (SiARC) (35 nm)/lithography resist (100 nm); (**b**) line/space pattern (pitch 97.5 ± 1 nm to 200 ± 1 nm with a step of 2.5 nm); (**c**) lines trimming with hydrofluoric (HF) 1% (critical dimension (CD) = 15 ± 1 nm); (**d**) neutral layer spin coating (**e**,**f**) lines removal with HF 1% and rinsing with propylene glycol methyl ether acetate (PGMEA); (**g**) selective grafting of polystyrene (PS); (**h**) directed self-assembly (DSA) of BCP L32.

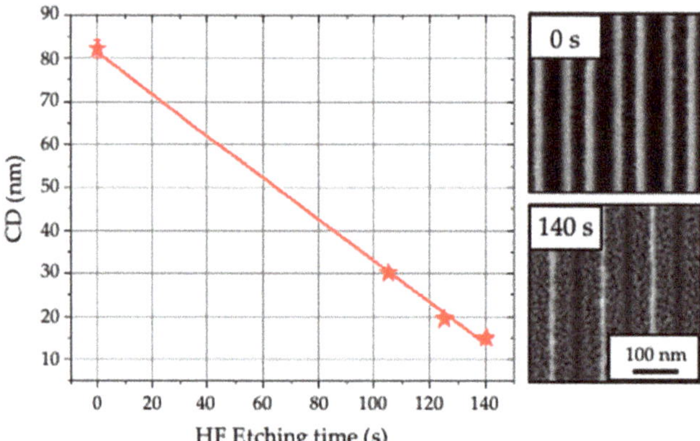

Figure 2. Lateral etching rate of TEOS lines with HF 1% and representative CD-SEM plan-view images of TEOS lines before etching (0 s) and after 140 s.

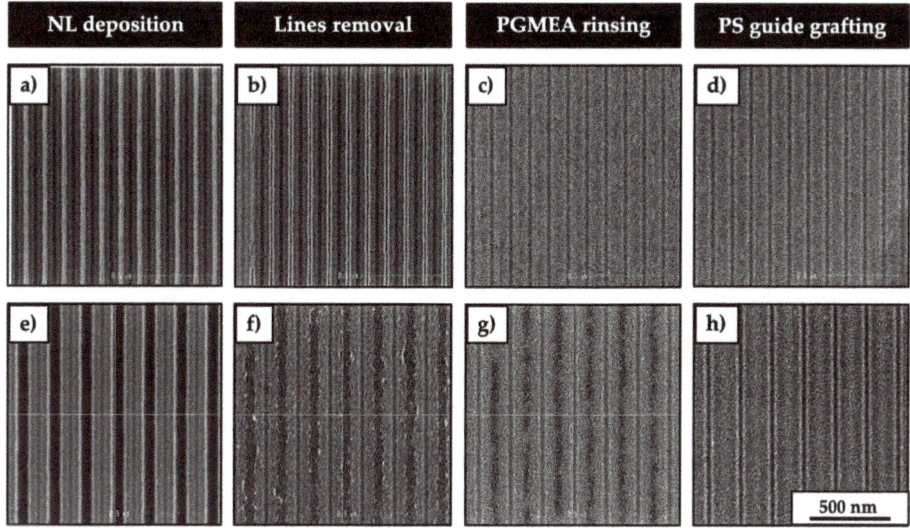

Figure 3. Representative plan view CD-SEM images of the steps: NLa deposition (**a**), NLb, NLc, NLd (**e**), lines removal with HF (**b**,**f**), PGMEA rinsing (**c**,**g**), and PS guide grafting (**d**,**h**).

The thicknesses of NLa, NLb, NLc, and NLd were adjusted optimizing the spin coating parameters. For NLa, the thermal treatment achieves the cross-linking of the polymeric chains inside the polymeric film while for NLb, NLc, and NLd, it promotes the grafting of the chains on the substrate. These experimental parameters are summarized in Table 1.

Table 1. Experimental parameters used for the neutral layer deposition.

Neutral Layer	Thickness Deposited (nm) (±0.5)	Thickness after Rinse (nm) (±0.5)	Thermal Treatment
NLa	7	7	250 °C/300 s
NLb	30	7	200 °C/75 s
NLc	15	4	200 °C/75 s
NLd	15	4	200 °C/75 s

Subsequently, the removal of the TEOS lines is mandatory to graft the PS and form the guidelines. In this context, the etching with HF 1% was optimized to 180 s to completely remove the TEOS lines (Figure S1, supplementary information). Here, it is possible to notice that in the case of NLa (Figure 3b), the guide left by the TEOS lines having the pitch of 120 nm is clearly visible and the edges are well defined. On the other hand, the guide left by the TEOS lines with the brush NLb, NLc, and NLd are not clearly visible in the pattern having a pitch of 120 nm. For this reason, the CD-SEM images are representative for the patterns having the pitch of 180 nm (Figure 3f). The next step consists in the PGMEA rinsing of the 300 mm wafer in order to remove the particles created during the etching of the TEOS lines and to remove the ungrafted polymer chains in the case of NL brushes. Figure 3c,g shows the effect of the PGMEA rinsing on the neutral layer. In the case of NLa (Figure 3c), the rinsing causes the shrinkage of the guides, making problematic the analysis with CD-SEM. For the brushes NLb, NLc, and NLd (Figure 3g) the guide is still visible, but the edges of the guides are not well defined. Finally, the grafting of the PS (Figure 3d,h) permits creating the guide necessary to have a chemo-epitaxy pattern for the subsequent DSA of the L32 BCP. The concentration of the PS solution and the spin-coating parameters were adjusted in order to obtain a thickness of 15 nm.

After the thermal treatment at 200 °C for 75 s and the subsequent PGMEA rinsing, the resulting PS-grafted layer thickness was around 4 nm. In both cases, the contrast is not enough to measure the CD of the PS guidelines, but the CD of these guidelines is defined by the CD of the TEOS lines previously measured by CD-SEM.

The last step of the trim-ox approach regards the deposition of the BCP L32 on the chemical pattern by using the Sokudo DUO track (Figure 1h). The concentration of the solution and the spin coating parameters were adjusted to obtain a BCP film thickness of 33 nm. Then, the wafers were thermally annealed at 240 °C for 900 s to achieve the self-assembly of the L32 film in perpendicularly oriented lamellae.

Figure 4 represents the top-view CD-SEM images of the self-assembled films for the four neutral layer employed (NLa, NLb, NLc, and NLd) for a TEOS lines pitch of 97.5 nm.

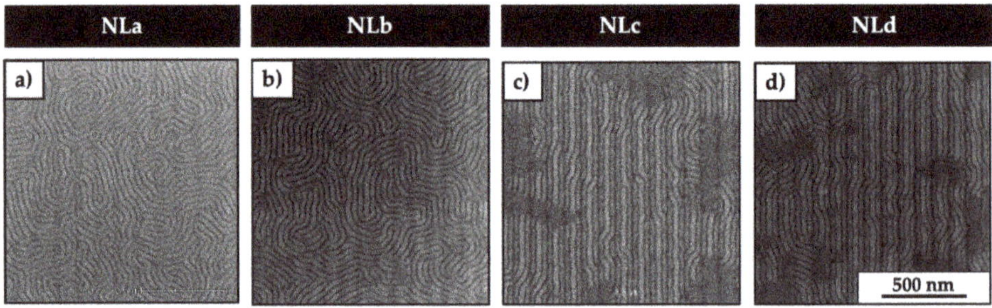

Figure 4. Representative top-view CD-SEM images of the self-assembled films when using NLa (**a**), NLb (**b**), NLc (**c**), and NLd (**d**) as neutral layers.

Here, it is possible to notice that in the case of NLa and NLb, as reported in Figure 4a,b, respectively, the chemical pattern does not guide the lamellae of the BCP, and fingerprint morphology was observed. On the other hand, when the neutral layers employed are the NLc and NLd, the perpendicular lamellae starts to be guided by the chemical pattern, and in these cases, the DSA takes place. This is probably due the low molar masses of NLc and NLd, which leads to the formation of a compact brush layer, and it avoids the PS molecules to penetrate inside the brush neutral layer. By the analysis of the CD-SEM images, the fraction of perpendicular lamellae for NLc and NLd was calculated to be about 77 ± 4% and 88.5 ± 4.4%, respectively. Considering the latest results, the NLd was selected as a reference neutral layer, and the influence of pitch dimensions for fixed CD = 15 nm of the TEOS lines were investigated, as shown in Figure 5.

Here, the dimension of the pitch varies from 97.5 ± 1 nm to 105 ± 1 nm corresponding to a MF from 3.05 to 3.28. The dimension of the pitch influences the perpendicular orientation with respect to the substrate (Figure 5a) and DSA (Figure 5b) of the BCP L32 lamellae. In the first case, it is possible to notice the presence of dark zones on the patterns corresponding to the parallel alignment of the BCP L32 lamellae with respect to the substrate. The area of these zones augments progressively passing from a pitch of 97.5 ± 1 to 105 ± 1 nm. In the latest, the alignment of the lamellae perpendicularly oriented passes from well-aligned (pitch 97.5 nm) to not aligned lamellae (pitch 105 nm) with respect to the chemical pattern pitch.

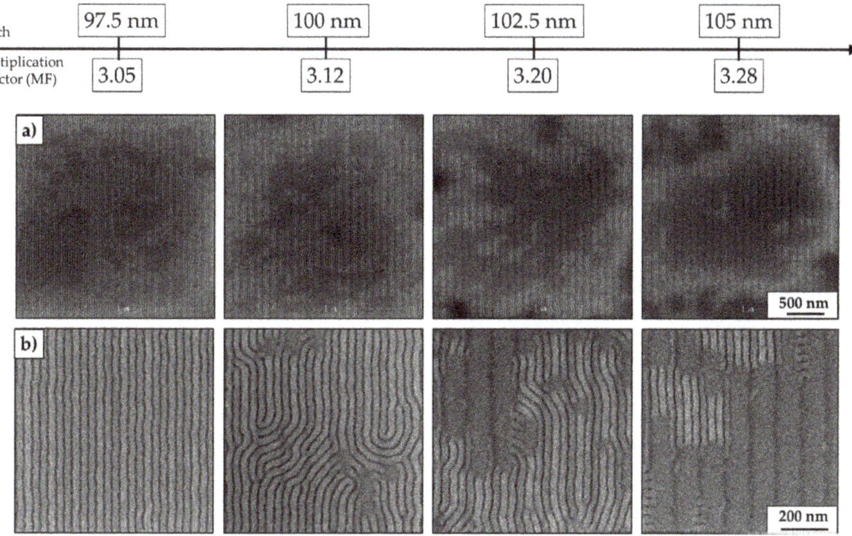

Figure 5. Representative CD-SEM images at low (**a**) et high (**b**) magnification, depicting the influence of the pitch dimension on the morphology of the BCP L32 for fixed CD ≈15 nm of TEOS lines. The multiplication factor (MF) is the result of Pitch/L_0, where L_0 = 32 nm is the period of L32 on a flat substrate.

Figure 6 reports the fractions of the perpendicularly oriented lamellae with respect to the substrate Δ (%) and the evolution of key metrics: period (L), Lines edge roughness (LER), Lines width roughness (LWR), and Herman's Orientational Parameter (P) as a function of TEOS lines pitch. These data are the results of the analysis of the CD-SEM images reported in Figure 5. The first graph from the top of the stack indicates the fraction of perpendicularly oriented lamellae (% Δ) as a function of the pitch. In the case, the fraction of lamellae perpendicularly oriented with respect to the substrate progressively decreases from 88.5 ± 4.4% to 28.9 ± 1.4% passing from a pitch of 97.5 ± 1 to 105 ± 1 nm of the TEOS lines.

The second graph of the stack in Figure 6 reports the evolution of L as a function of the pitch. Here, it is possible to notice that L remain constant around values of ≈32 ± 2 nm independently of the pitch of the TEOS lines employed.

The third graph of each stack reports the LER (3σ) and LWR (3σ) evolution as a function of the pitch considering the systems with lamellae perpendicularly oriented with respect to the substrate. LER and LWR represent the deviation from a straight-line edge and the deviation from a uniform line width, respectively [19,21]. In this case, the LER and LWR values fluctuate in the range of LER ≈2.7–4.2 nm and LWR ≈4.8–7 nm. In particular, for the pitch of 97.5 ± 1 and 100 ± 1 nm, respectively, these values remain constant with nominal values of LER ≈2.7 nm and LWR ≈4.8 nm. For a pitch of 102.5 nm, the LER and LWR tend to increase to values of ≈4.2 nm and ≈7 nm, respectively. Finally, for a pitch of 105 nm, the LER and LWR decrease, reaching values of 3.6 nm and 6.2 nm, respectively. This morphological evolution is characteristic of non-commensurability between the chemical pattern and the period of the BCP [9].

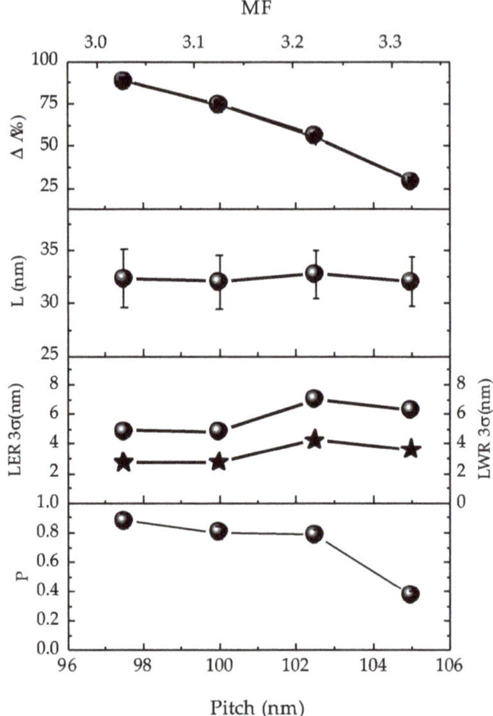

Figure 6. Evolution of Δ (%), period L, line edge roughness (LER) (stars), line width roughness (LWR) (spheres), and Herman's orientational (P) parameter as a function of the TEOS pitch size and multiplication factor (MF) in L32 BCP thin films.

The fourth graph of each stack reports the Herman's Orientational Parameter (P) evolution as a function of TEOS pitch. The P parameters [22,23] gives a measure of the uniformity of the lamellae within an image. In this case, the more the P parameter tends to 1, the more the lamellae are oriented vertically following the chemical patterns created on the substrate [19,23].

The P values of the perpendicular lamellae decrease with the TEOS pitch size passing from $P = 0.879 \pm 0.044$ for a pitch of 97.5 ± 1 nm to $P = 0.378 \pm 0.019$ for a pitch of 105 ± 1 nm. These values of P indicate a progressive loss of the DSA with the increasing of the pitch size because the P parameter tends to move away from the values of $P = 1$, which indicates the perfect ordering of the lamellae nanostructures.

4. Discussions

The DSA of the lamella forming PS-*b*-PMMA is affected by the nature of the neutral layers employed. In the case of cross-linkable (NLa) and highest molar mass brush (NLb) neutral layers, the chemical contrast between the PS guidelines and the NL background is not enough to drive the DSA of the BCP. In the first case, this is due to the shrinkage of the NLa after the PGMEA rinsing that limits the possibility of having the PS guideline (Figure 3b,c). In the second case, this is probably due to the higher molar mass of the brush neutral layer compared to the molar mass of the PS that forms the guideline. This is probably because the interface between the polymeric chains of the PS guideline and the NL brushes is not clear-cut because of an interpenetration between the polymeric chains of NL and PS that does not permit obtaining the DSA of the perpendicular lamellae. In fact, for NLc and NLd, the neutral layers have molar mass comparable to the one of the polymer that forms the PS guideline. In this case, the chemical pattern created guides the perpendicular-oriented lamellae, achieving the

DSA. As previously reported in the literature, experimental [24,25] and computational [26] studies confirmed that the interpenetration region between systems consisting of polymeric films deposited onto grafted brush layers increases with the molar mass of the brush layer by fixing the molar mass of the polymer film. Although these previous studies investigated the interpenetration between layers and not the lateral interpenetration, they can be useful to give a plausible explanation to the results obtained in the present work.

The overall picture of the reported data demonstrate that the PS guideline of the chemo-epitaxial pattern can be obtained exploiting the trimming and the following etching of TEOS L/S patterns without the use of polymeric pinning materials, which is commonly exploited in the existing DSA methods.

In this context, considering the SMART process, the photoresist is deposited, etched, and stripped on top of the cross-linked PS-*r*-PMMA layer to form the guidelines. The critical dimension (CD) of the guidelines is defined by the photolithographic methods. Therefore, this introduces limitations in terms of the CD range that is accessible. Moreover, the etching process to transfer the pattern into the neutral layer and the stripping process to remove the photoresist could cause modifications in the surface properties of the resulting neutral patterns. In a similar way, the LiNe process fabricated the guideline, but in this case, the limits of the CD were overcome by means of the O_2 plasma trimming process of the resist to reduce its CD. On the other side, the trimming process introduces changing in the affinity with respect to one block of the BCP. The COOL process is very similar to the LiNe process, but the guideline is represented by the resist that is slimmed by means of the etching process. Thus, there no need to strip the resist during the process. Nevertheless, the affinity of the resist guideline with one of the BCP blocks is guaranteed by the etching process, and these can introduce limitations regarding its univocal affinity with one of the BCP blocks.

According to the International Technology Roadmap for Semiconductor (ITRS) [27] and the International Roadmap for Devices and Systems (IRDS) [28], the LER and LWR of nanostructures that have dimensions below 15 nm are requested to be lower than 0.9 and 1.2 nm, respectively, because higher values could afflict the performance of the final microelectronics devices. In particular, high values are deleterious for circuit elements. For transistor gate structures, the roughness causes significant variations in the off-current, as well as affecting threshold voltages. For nanometer-scale interconnects, the roughness increases both resistance and capacitance [21,29–31]. Table 2 reports the LER and LWR literature values for lamellae PS-*b*-PMMA nanostructures having a period L comparable to the one reported in the following work by a different DSA approach. Although the experimental LER and LWR values are almost 2–3 times higher than the limit values imposed by ITRS and IRDS, it is possible to notice that the LER and LWR values obtained with the Trim-Ox approach presented in this work are in line with the literature data of the most studied approach. Consequently, the Trim-Ox approach could be considered as a competitor method for the integration of the DSA in the existing conventional photolithography.

Table 2. LER and LWR values for various approaches that include LiNe, SMART, COOL, and the Trim-Ox approach presented in the following work.

DSA Approach	LER (3σ) (nm)	LWR (3σ) (nm)
LiNe [32–35]	2.49–3.12	2.45–4.0
SMART [36,37]	3.5–4.0	2.9–5.0
COOL [17,38]	3.5–4.5	4.0–6.0
Trim-Ox	2.7–4.2	4.8–7

5. Conclusions

The Trim-Ox approach was implemented in order to obtain the DSA of BCP L32 having an intrinsic period of 32 nm. It has demonstrated the possibility to finely tune the CD dimension of the TEOS line with a simple etching in HF 1%, reaching the CD = 15 nm. Moreover, it has been demonstrated that the integration of chemical patterns composed of grafted NLd and PS guide on a 300 mm wafer can be achieved exploiting the TEOS lines. The process developed is fully compatible with 300 mm clean-room facilities, and the performance in terms of roughness can be considered at the same level to the state-of-the-art methods reported in the literature.

6. Patents

Patent related to this work is pending, application number FR1911542.

Supplementary Materials: The following are available online at http://www.mdpi.com/2079-4991/10/12/2443/s1, Figure S1: Evolution of the TEOS lines etching as function of the HF etching time.

Author Contributions: The author's individual contributions are the following: conceptualization, T.J.G., A.G., A.P. and R.T.; methodology, T.J.G.; formal analysis, T.J.G.; writing—original draft preparation, T.J.G.; supervision, R.T.; project administration, R.T.; funding acquisition, R.T. and P.N. All authors have read and agreed to the published version of the manuscript.

Funding: This research was partly funded by the DiSABloc project of the Grenoble Nanoscience Foundation.

Acknowledgments: Christophe Navarro, Célia Nicolet from Arkema and Kaumba Sakavuyid from Brewer Science for the synthesis of the polymers. Fabienne Guennou, Aurélie Le Pennec, Maxime Argoud, Guido Rademaker from LETI for fruitful discussions.

Conflicts of Interest: The authors declare no conflict of interest.

References

1. Rahman, A.; Ashraf, A.; Xin, H.; Tong, X.; Sutter, P.; Eisaman, M.D.; Black, C.T. Sub-50-nm self-assembled nanotextures for enhanced broadband antireflection in silicon solar cells. *Nat. Commun.* **2015**, *6*, 5963. [CrossRef] [PubMed]

2. Stefik, M.; Guldin, S.; Vignolini, S.; Wiesner, U.; Steiner, U. Block copolymer self-assembly for nanophotonics. *Chem. Soc. Rev.* **2015**, *44*, 5076–5091. [CrossRef]

3. Yi, H.; Bao, X.Y.; Tiberio, R.; Philip Wong, H.S. A General Design Strategy for Block Copolymer Directed Self-Assembly Patterning of Integrated Circuits Contact Holes using an Alphabet Approach. *Nano Lett.* **2015**, *15*, 805–812. [CrossRef] [PubMed]

4. Yu, H.; Qiu, X.; Moreno, N.; Ma, Z.; Calo, V.M.; Nunes, S.P.; Peinemann, K.V. Self-Assembled Asymmetric Block Copolymer Membranes: Bridging the Gap from Ultra- to Nanofiltration. *Angew. Chem. Int. Ed.* **2015**, *54*, 13937–13941. [CrossRef]

5. Bates, F.S. Polymer-Polymer Phase Behavior. *Science* **1991**, *251*, 898–905. [CrossRef] [PubMed]

6. Lo, T.-Y.; Krishnan, M.R.; Lu, K.-Y.; Ho, R.-M. Silicon-containing block copolymers for lithographic applications. *Prog. Polym. Sci.* **2017**. [CrossRef]

7. Giammaria, T.J.; Ferrarese Lupi, F.; Seguini, G.; Perego, M.; Vita, F.; Francescangeli, O.; Wenning, B.; Ober, C.K.; Sparnacci, K.; Antonioli, D.; et al. Micrometer-Scale Ordering of Silicon-Containing Block Copolymer Thin Films via High-Temperature Thermal Treatments. *ACS Appl. Mater. Interfaces* **2016**, *8*, 9897–9908. [CrossRef]

8. Giammaria, T.J.; Laus, M.; Perego, M. Technological strategies for self-assembly of ps-b-pdms in cylindrical sub-10 nm nanostructures for lithographic applications. *Adv. Phys. X* **2018**, *3*, 391–411. [CrossRef]

9. Ji, S.; Wan, L.; Liu, C.-C.; Nealey, P.F. Directed self-assembly of block copolymers on chemical patterns: A platform for nanofabrication. *Prog. Polym. Sci.* **2016**, *54–55*, 76–127. [CrossRef]

10. Tiron, R.; Chevalier, X.; Couderc, C.; Pradelles, J.; Bustos, J.; Pain, L.; Navarro, C.; Magnet, S.; Fleury, G.; Hadziioannou, G. Optimization of block copolymer self-assembly through graphoepitaxy: A defectivity study. *J. Vac. Sci. Technol. B* **2011**, *29*, 06F206. [CrossRef]

11. Cheng, J.; Doerk, G.S.; Rettner, C.T.; Singh, G.; Tjio, M.; Truong, H.; Arellano, N.; Balakrishnan, S.; Brink, M.; Tsai, H.; et al. Customization and design of directed self-assembly using hybrid prepatterns. *Proc. SPIE 9423 Altern. Lithogr. Technol. VII* **2015**, 942307. [CrossRef]

12. Cheng, J.Y.; Sanders, D.P.; Truong, H.D.; Harrer, S.; Friz, A.; Holmes, S.; Colburn, M.; Hinsberg, W.D. Simple and Versatile Methods To Integrate Directed Self-Assembly with Optical Lithography Using a Polarity-Switched Photoresist. *ACS Nano* **2010**, *4*, 4815–4823. [CrossRef] [PubMed]

13. Wan, L.; Ruiz, R.; Gao, H.; Albrecht, T.R. Self-Registered Self-Assembly of Block Copolymers. *ACS Nano* **2017**, *11*, 7666–7673. [CrossRef] [PubMed]

14. Cushen, J.; Wan, L.; Blachut, G.; Maher, M.J.; Albrecht, T.R.; Ellison, C.J.; Willson, C.G.; Ruiz, R. Double-Patterned Sidewall Directed Self-Assembly and Pattern Transfer of Sub-10 nm PTMSS-b-PMOST. *ACS Appl. Mater. Interfaces* **2015**, *7*, 13476–13483. [CrossRef] [PubMed]

15. Liu, C.-C.; Ramírez-Hernández, A.; Han, E.; Craig, G.S.W.; Tada, Y.; Yoshida, H.; Kang, H.; Ji, S.; Gopalan, P.; de Pablo, J.J.; et al. Chemical Patterns for Directed Self-Assembly of Lamellae-Forming Block Copolymers with Density Multiplication of Features. *Macromolecules* **2013**, *46*, 1415–1424. [CrossRef]

16. Kim, J.; Wan, J.; Miyazaki, S.; Yin, J.; Cao, Y.; Her, Y.J.; Wu, H.; Shan, J.; Kurosawa, K.; Lin, G. The SMART^{TM} process for directed block co-polymer self-assembly. *J. Photopolym. Sci. Technol.* **2013**, *26*, 573–579. [CrossRef]

17. Seino, Y.; Kasahara, Y.; Sato, H.; Kobayashi, K.; Miyagi, K.; Minegishi, S.; Kodera, K.; Kanai, H.; Tobana, T.; Kihara, N.; et al. A novel simple sub-15nm line-and-space patterning process flow using directed self-assembly technology. *Microelectron. Eng.* **2015**, *134*, 27–32. [CrossRef]

18. Paquet, A.; Le Pennec, A.; Gharbi, A.; Giammaria, T.J.; Rademaker, G.; Pourteau, M.-L.; Mariolle, D.; Navarro, C.; Nicolet, C.; Chevalier, X.; et al. Spacer patterning lithography as a new process to induce block copolymer alignment by chemo-epitaxy. *Proc. SPIE 10958 Nov. Patterning Technol. Semicond. MEMS/NEMS MOEMS* **2019**, 109580M. [CrossRef]

19. Murphy, J.N.; Harris, K.D.; Buriak, J.M. Automated Defect and Correlation Length Analysis of Block Copolymer Thin Film Nanopatterns. *PLoS ONE* **2015**, *10*, e0133088. [CrossRef]

20. Spierings, G.A.C.M. Wet chemical etching of silicate glasses in hydrofluoric acid based solutions. *J. Mater. Sci.* **1993**, *28*, 6261–6273. [CrossRef]

21. Kim, H.W.; Lee, J.Y.; Shin, J.; Woo, S.G.; Cho, H.K.; Moon, J.T. Experimental investigation of the impact of LWR on sub-100-nm device performance. *IEEE Trans. Electron. Devices* **2004**, *51*, 1984–1988. [CrossRef]

22. Qiang, Z.; Zhang, L.; Stein, G.E.; Cavicchi, K.A.; Vogt, B.D. Unidirectional Alignment of Block Copolymer Films Induced by Expansion of a Permeable Elastomer during Solvent Vapor Annealing. *Macromolecules* **2014**, *47*, 1109–1116. [CrossRef]

23. Jung, Y.S.; Ross, C.A. Orientation-controlled self-assembled nanolithography using a polystyrene—Polydimethylsiloxane block copolymer. *Nano Lett.* **2007**, *7*, 2046–2050. [CrossRef] [PubMed]

24. Sparnacci, K.; Chiarcos, R.; Gianotti, V.; Laus, M.; Giammaria, T.J.; Perego, M.; Munaò, G.; Milano, G.; De Nicola, A.; Haese, M.; et al. Effect of Trapped Solvent on the Interface between PS-b-PMMA Thin Films and P(S-r-MMA) Brush Layers. *ACS Appl. Mater. Interfaces* **2020**, *12*, 7777–7787. [CrossRef] [PubMed]

25. Lee, H.; Jo, S.; Hirata, T.; Yamada, N.L.; Tanaka, K.; Kim, E.; Yeol, D. Interpenetration of chemically identical polymer onto grafted substrates. *Polymer* **2015**, *74*, 70–75. [CrossRef]

26. Gay, C. Wetting of a Polymer Brush by a Chemically Identical Polymer Melt. *Macromolecules* **1997**, *30*, 5939–5943. [CrossRef]

27. The International Technology Roadmap for Semiconductors 2.0: Lithography, 2015. Available online: http://www.itrs2.net/itrs-reports.html (accessed on 1 November 2020).

28. International Roadmap for Devices and Systems 2017 Edition: Lithography, 2017. Available online: https://irds.ieee.org/editions/2017 (accessed on 1 November 2020).

29. Asenov, A.; Kaya, S.; Brown, A.R. Intrinsic parameter fluctuations in decananometer MOSFETs introduced by gate line edge roughness. *IEEE Trans. Electron. Devices* **2003**, *50*, 1254–1260. [CrossRef]

30. Steinhögl, W.; Schindler, G.; Steinlesberger, G.; Traving, M.; Engelhardt, M. Impact of line edge roughness on the resistivity of nanometer-scale interconnects. *Microelectron. Eng.* **2004**, *76*, 126–130. [CrossRef]

31. Stucchi, M.; Bamal, M.; Maex, K. Impact of line-edge roughness on resistance and capacitance of scaled interconnects. *Microelectron. Eng.* **2007**, *84*, 2733–2737. [CrossRef]

32. Seidel, R.; Williamson, L.; Her, Y.; Kim, J.; Lin, G.; Nealey, P.; Gronheid, R. The role of guide stripe chemistry in block copolymer directed self-assembly. *Adv. Patterning Mater. Process. XXXII* **2015**, *9425*, 94250W. [CrossRef]

33. Sunday, D.F.; Chen, X.; Albrecht, T.R.; Nowak, D.; Rincon Delgadillo, P.; Dazai, T.; Miyagi, K.; Maehashi, T.; Yamazaki, A.; Nealey, P.F.; et al. Influence of Additives on the Interfacial Width and Line Edge Roughness in Block Copolymer Lithography. *Chem. Mater.* **2020**, *32*, 2399–2407. [CrossRef]

34. Chan, B.T.; Tahara, S.; Parnell, D.; Rincon Delgadillo, P.A.; Gronheid, R.; de Marneffe, J.-F.; Xu, K.; Nishimura, E.; Boullart, W. 28nm pitch of line/space pattern transfer into silicon substrates with chemo-epitaxy Directed Self-Assembly (DSA) process flow. *Microelectron. Eng.* **2014**, *123*, 180–186. [CrossRef]

35. Patel, K.; Wallow, T.; Levinson, H.J.; Spanos, C.J. Comparative study of line width roughness (LWR) in next-generation lithography (NGL) processes. *Proc. SPIE 7640 Opt. Microlithogr. XXIII* **2010**, 76400T. [CrossRef]

36. Kim, J.; Yin, J.; Cao, Y.; Her, Y.; Petermann, C.; Wu, H.; Shan, J.; Tsutsumi, T.; Lin, G. Toward high-performance quality meeting IC device manufacturing requirements with AZ SMART DSA process. *Altern. Lithogr. Technol. VII* **2015**, *9423*, 94230R. [CrossRef]

37. Somervell, M.; Yamauchi, T.; Okada, S.; Tomita, T.; Nishi, T.; Kawakami, S.; Muramatsu, M.; Iijima, E.; Rastogi, V.; Nakano, T.; et al. Driving DSA into volume manufacturing. *Proc. SPIE 9425 Adv. Patterning Mater. Process. XXXII* **2015**, 94250Q. [CrossRef]

38. Seino, Y.; Kasahara, Y.; Sato, H.; Kobayashi, K.; Kubota, H.; Minegishi, S.; Miyagi, K.; Kanai, H.; Kodera, K.; Kihara, N.; et al. Directed self-assembly lithography using coordinated line epitaxy (COOL) process. *Altern. Lithogr. Technol. VII* **2015**, *9423*, 942316. [CrossRef]

Publisher's Note: MDPI stays neutral with regard to jurisdictional claims in published maps and institutional affiliations.

 *nanomaterials* 

Article

Directed Self-Assembly of Polystyrene Nanospheres by Direct Laser-Writing Lithography

Eleonora Cara [1,2], **Federico Ferrarese Lupi** [1,*], **Matteo Fretto** [1], **Natascia De Leo** [1], **Mauro Tortello** [2], **Renato Gonnelli** [2], **Katia Sparnacci** [3] **and Luca Boarino** [1]

[1]   Advanced Materials Metrology and Life Sciences Division, Istituto Nazionale di Ricerca Metrologica, Strada delle Cacce 91, 10135 Torino, Italy; e.cara@inrim.it (E.C.); m.fretto@inrim.it (M.F.); n.deleo@inrim.it (N.D.L.); l.boarino@inrim.it (L.B.)

[2]   Dipartimento di Scienza Applicata e Tecnologia (DISAT), Politecnico di Torino, Corso Duca degli Abruzzi 24, 10129 Torino, Italy; mauro.tortello@polito.it (M.T.); renato.gonnelli@polito.it (R.G.)

[3]   Dipartimento di Scienze e Innovazione Tecnologica (DIST), Universitá del Piemonte Orientale "A. Avogadro", INSTM, Viale T. Michel 11, 15121 Alessandria, Italy; katia.sparnacci@uniupo.it

*   Correspondence: f.ferrareselupi@inrim.it; Tel.: +39-011-3919-449

Received: 24 December 2019; Accepted: 31 January 2020; Published: 7 February 2020

Abstract: In this work, we performed a systematic study on the effect of the geometry of pre-patterned templates and spin-coating conditions on the self-assembling process of colloidal nanospheres. To achieve this goal, large-scale templates, with different size and shape, were generated by direct laser-writer lithography over square millimetre areas. When deposited over patterned templates, the ordering dynamics of the self-assembled nanospheres exhibits an inverse trend with respect to that observed for the maximisation of the correlation length ξ on a flat surface. Furthermore, the self-assembly process was found to be strongly dependent on the height (H) of the template sidewalls. In particular, we observed that, when H is 0.6 times the nanospheres diameter and spinning speed 2500 rpm, the formation of a confined and well ordered monolayer is promoted. To unveil the defects generation inside the templates, a systematic assessment of the directed self-assembly quality was performed by a novel method based on Delaunay triangulation. As a result of this study, we found that, in the best deposition conditions, the self-assembly process leads to well-ordered monolayer that extended for tens of micrometres within the linear templates, where 96.2% of them is aligned with the template sidewalls.

Keywords: directed self-assembly; nanospheres lithography; colloidal nanospheres; direct laser-writing

---

1. Introduction

Nanospheres lithography (NSL) is a manufacturing technique based on the self-assembly (SA) process of colloidal spheres [1]. Monodisperse suspensions of polystyrene (PS) nanospheres (NSs) deposited on a substrate form colloidal crystals consisting in single or multiple layers, exhibiting hexagonal close-packed (HCP) symmetry. In the last decades, NSL gained increasing attention in nanotechnology due to the possibility to realise several periodic patterns over large area and at reasonable cost, including photonic structures [2] or devices for nanoelectronics [3] and plasmonics [4]. However, the SA process exhibits intrinsic variability, resulting in the generation of lattice defects and the formation of multiple domains. These irregularities hinder advanced applications in which precise spatial positioning of the nanostructures is required. In this context, the design of an experimental procedure with stable output is demanded for the fabrication of well-ordered single domains with controlled size and regular shape.

An interesting solution to overcome this limitation is represented by the use of substrate modifications to aid the formation of single-layered crystals of NSs. This approach, also called directed self-assembly (DSA), has been successfully proposed for other self-assembling systems, such as Block Copolymers (BCPs), receiving great consideration so far due to its wide applicability in key technological sectors such as microelectronics [5–7]. The substrate can be modified either by a chemical [8,9] or topographic templates generated prior the SA process [10]. In the latter case, the bottom-up SA process is directed by the presence of confining structures such as linear or circular gratings, defined by conventional top-down lithographic approaches [11]. The geometrical dimensions of the topographic templates can be tailored to be commensurate to the characteristic dimensions of the SA material (e.g., diameter of the NSs or center-to-center distance for BCPs).

The development of DSA processes applied to NSL has been mainly dedicated to the confinement of few NSs [12–14] or to achieve size separation of polydispersed NSs [15]. The present work aims to extend the DSA process over large area, to allow the formation of single-grain domains highly oriented inside pre-patterned templates throughout a several square millimetre area. To meet this objective, direct laser writing (DLW) lithography and reactive ion etching (RIE) were combined to fabricate micrometric templates with different shapes and sizes. The deposition of the NSs in the templates was performed by spin coating, and the dynamic parameters were varied starting from the insights of our previous work [16]. In particular, we investigate the formation of the NSs monolayer through the analysis of the confinement and the ordering processes. The former was carried out by means of atomic force microscopy (AFM) and scanning electron microscopy (SEM), whereas the NSs ordering was evaluated through an image-processing method measuring the domains orientation. These analyses contribute to increase the repeatability of NSL and expand its applicability through DSA to address the necessities of the development of novel devices for photonics [17], chemical sensing [18,19], data storage [20], and optoelectronics [21].

2. Materials and Methods

2.1. Direct Laser-Writing Patterning

The DLW lithography (Heidelberg μPG101 laser writer, Heidelberg, Germany) was performed on polished silicon wafers (MEMC Electronic Materials, Novara, Italy) covered by a thermal oxide layer with thickness ranging between 50 nm and 200 nm. An optical resist (AZ 1505 Merck Performance Materials GmbH, Darmstadt, Germany) was deposited over the SiO_2 substrate (Figure 1a) and exposed with a laser beam ($\lambda = 375$ nm, diameter of 800 nm and intensity of 10 mW). The resist was afterward developed for 40 s in a 1:1 solution of the developer (AZ Developer Merck Performance Materials GmbH) and H_2O. The resulting pattern left the SiO_2 layer exposed, as shown in Figure 1b. The templates sizes were designed to confine an integer number of NSs, including hexagonal templates with a diagonal length of 4.75 μm and linear ones with a width of 3 μm and length of 200 μm, while Figure 1 only reports the hexagonal configuration as an example.

2.2. Template Fabrication

The DLW pattern was transferred to the oxide layer by reactive ion etching process (RIE) (Figure 1c). The chemically reactive plasma was obtained by mixing CHF_3 and Ar with a flow ratio of 54 sccm to 29 sccm. The plasma was generated with a residual pressure of 180 Pa and an applied RF power of 300 W, with a typical reflected power of 25 W. Under these operating conditions, the etching rate on SiO_2 was 10 nm min^{-1} and the time was selected to reach different depths. After the etching step, the excess resist was removed with acetone and the final patterned substrate was characterised by a non-contact 3D surface profiler (Sensofar S Neox, Barcelona, Spain) (Figure 1e) and a field emission gun (FEG) SEM (FEI Inspect-FTM, Hillsboro, OR, USA) (Figure 1f).

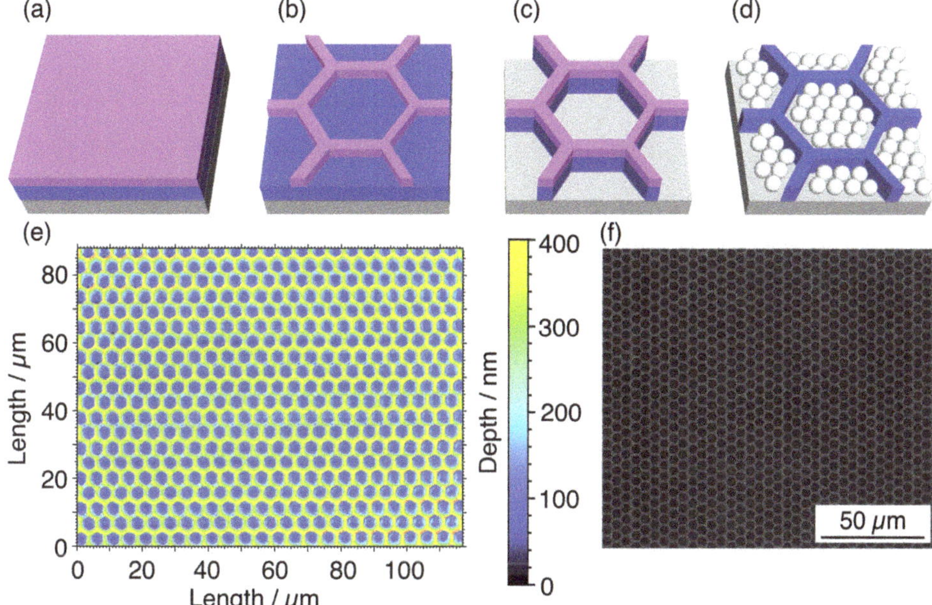

Figure 1. (**a**) The silicon (grey) substrate with a SiO_2 layer (blue) is covered with a layer of photosensitive resist (purple). (**b**) The hexagonal pattern is designed on the resist by direct laser writing (DLW) lithography and developed to expose the underlying SiO_2 layer, (**c**) which is then etched by reactive ion etching process (RIE). (**d**) The PS NSs are deposited inside the resulting template. (**e**) Optical profilometry image and (**f**) SEM micrograph at low magnification showing the hexagonal templates over large area after the etching.

2.3. Nanospheres Deposition

Despite the other NSs deposition methods that have been proposed so far, such as doctor blading [22] or Langmuir–Blodgett coating [23], in this work, we used the spin coating technique. Such choice was motivated by the aim to develop proper protocols to promote the applicability of DSA processing in industrial nanomanufacturing already relying on this method. The patterned substrates were cleaned in an ultrasonic bath of acetone and isopropyl alcohol. The surface was treated by O_2 plasma for 6 minutes at 40 W with a residual pressure of 3 Pa to make it hydrophilic. The PS NSs were synthesised using the emulsion polymerisation of styrene using sodium dodecyl sulfate as surfactant and potassium persulfate as the initiator [19]. The NSs had diameter equal to (250 ± 4) nm and presented negative charges at the surface, due to the decomposition of the initiator, thus stabilising the aqueous suspension against aggregation. We drop coated all the samples with 60 µL of the suspension and spread it by spin coating (WS-400B-6NPP/LITE Laurell Technologies, North Wales, PA, USA) in two steps. In the first step, we set the speed and acceleration to 500 rpm and 410 rpm/s, respectively, and the duration to 10 s. For the second step, we modified the spinning speed to test the confinement process while keeping the duration at 30 s. An illustration of the result is shown in Figure 1d.

2.4. SEM Characterisation and Image Processing

The characterisation of NSs self-assembly inside the templates was performed by a systematic analysis of the SEM micrographs. We set conditions for the SEM imaging with $V = 10$ kV, planar configuration at the optimum working distance of 10 mm and magnification of 10,000. For a quantitative analysis of the DSA process, we processed the images by means of a MATLAB routine which operates by recognising the NSs inside the templates and by mapping the lattice according to Delaunay triangulation. Then, it identifies deviations from the ideal HCP lattice by counting the number of nearest neighbours to each particle. The orientation of all the unit HCP cells is extrapolated with an angular resolution of 1° in the range of possible orientations of the crystals between −30° and 30°. A complete description of the operating principle of the software is reported in reference 16.

2.5. Atomic Force Microscopy Characterisation

The surface topography on the NSs soft material was investigated by means of atomic force microscopy (Bruker Corp. INNOVA microscope) by using etched Si probes (Bruker RTESPA-300, Billerica, MA, USA) with nominal spring constant of $40\,N\,m^{-1}$ and tip radius of $8\,nm$. The measurements were performed in tapping mode with a resonance frequency of $230\,kHz$ and scanning rate of $0.5\,Hz$. The analysis of the AFM micrographs was carried out by the freeware Gwyddion. The plane inclination was corrected by fitting a plane through three points on the optically flat SiO_2 mesas and by setting the scale zero position at the same level.

3. Results and Discussion

3.1. Nanospheres Ordering

The deposition of NSs over the patterned substrates was performed by spin-coating process. We set the spinning speed and acceleration to $1250\,rpm$ and $410\,rpm/s$, in agreement to our previous experiments focused on the maximisation of the degree of order, expressed in terms of correlation length ξ [16], on flat unpatterned substrates. Figure 2a shows a low-magnification SEM micrograph of both flat and patterned areas on the substrate. On the flat portion of the sample, the formation of large grains is preserved, as highlighted in Figure 2b by the overlapped colour map. Each coloured region corresponds to a grain or domain in which the orientation of the HCP lattice is uniform, whereas it varies randomly in the neighbouring domains separated by the grain boundaries. However, the same spinning conditions were found inadequate for the SA inside the templates, leading to the accumulation of NSs in multiple layers reported in Figure 2c.

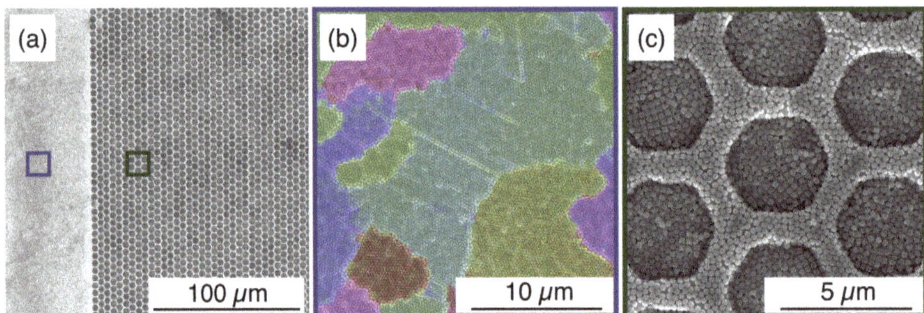

Figure 2. (**a**) Low-magnification SEM micrograph preliminarily comparing the SA of NSs on the flat and patterned area of the substrate. The blue and green frames correspond to the high-magnification SEM image (**b**) of the self-assembled monolayer, where the domains are highlighted in different colours showing a random change in the orientation from one to the other, and (**c**) inside the templates, where multiple layers are formed.

This preliminary result highlights the differences of the SA induced on a flat substrate and inside the templates. The SA process has been described in literature as the interaction of capillary forces between two adjacent NSs, responsible for the hexagonal packing. In the presence of a geometrical constraint, such capillary forces also act across the edges of the templates, which introduce a perturbation of the conventional SA process [13,24,25]. To quantify the effect of the perturbation on the long-range ordering and to optimise the confinement of the NSs, we realised a new set of samples by varying both the height of the sidewalls and the spinning conditions. In particular, the spinning speed were set to 1250 rpm, 2000 rpm and 2500 rpm, whereas the selected heights H were 50 nm (H = 0.2·D), 100 nm (H = 0.4·D), 150 nm (H = 0.6·D) and 200 nm (H = 0.8·D). The maximum value of H (i.e., 200 nm) was chosen below the NSs diameter since excessive height would result in a physical barrier promoting the stratification in multiple layers. Figure 3 reports a tabular comparison of the SEM micrographs of the colloidal crystal, where the sidewalls height and spinning speed are varied along the columns and rows, respectively. The structures were patterned with hexagonal shape for its similarity to the characteristic packing symmetry of the NSs.

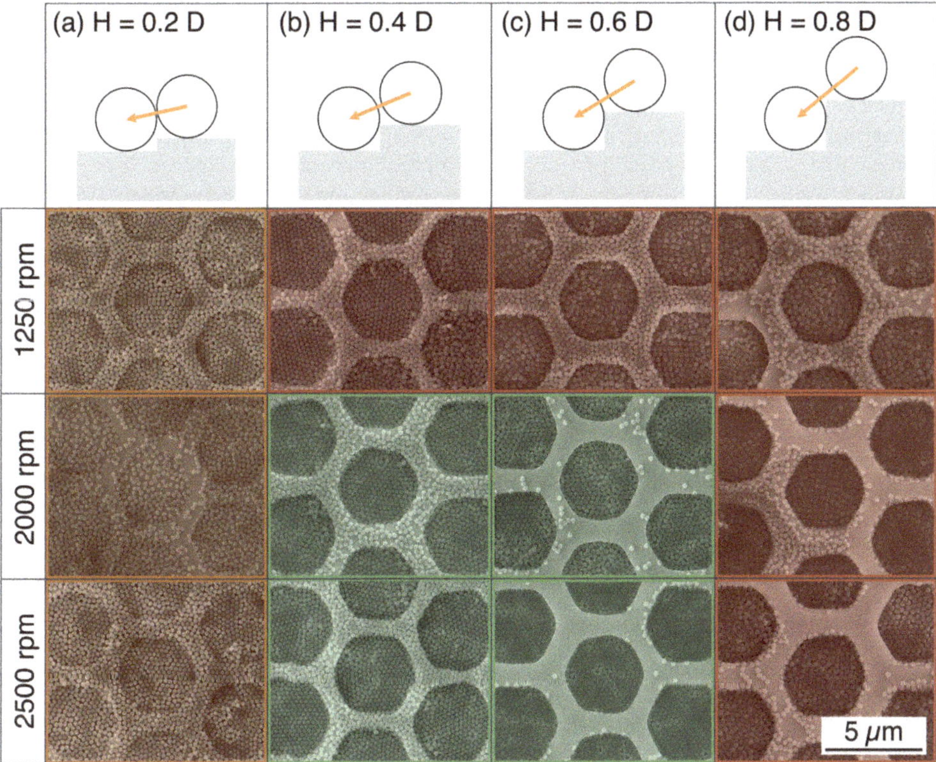

Figure 3. Tabular comparison of the SEM images of NSs assemblies inside hexagonal templates. The height of the confining wall varies along the columns, column (**a**) H = 0.2·D, column (**b**) H = 0.4·D, column (**c**) H = 0.6·D and column (**d**) H = 0.8·D. The spin-coating speed is varied in the rows from 1250 rpm to 2500 rpm. The SEM images highlighted in red and orange refer to unsuitable confinement conditions, whereas those coloured in green display suitable confinement.

In the case of templates with H = 0.2·D (i.e., depth of 50 nm) shown in Figure 3a, the NSs self-assemble into monolayers irrespective on spinning speed. However, under these particular conditions, the orientation of the domains is not influenced by the presence of the template, as testified by the formation of grains with same orientation across the edges. For this reason, these conditions are not proper for NSs confinement and the corresponding images are coloured in orange.

On the contrary, in templates with H = 0.4·D and H = 0.6·D in Figure 3b,c, the arrangement of the NSs presents a marked dependence on the spinning parameters. For depositions performed at 1250 rpm, we observed the formation of multiple layers inside the templates (red images in Figure 3), preventing the lithographic use of the confined NSs. Such an issue can be solved by increasing the spinning speed to 2000 rpm and 2500 rpm. In this case, the NSs arrange in a single layer confined inside the templates and, despite the presence of residual NSs on the mesas in between adjacent templates, no domains are continuously ordered across the edges. In these conditions, the formation of the monolayer is facilitated and visibly influenced by the presence of the templates, the corresponding micrographs are coloured in green in Figure 3. Finally, when deposited in templates with H = 0.8·D (i.e., depth of 200 nm), the NSs accumulate in multiple layers independently of the spin-coating speed so that these conditions are not suitable for lithographic purposes (SEM images coloured in red in Figure 3d). In light of this result, the structures with H/D ratios of 0.2 and 0.8 seems to be either too shallow or too deep to produce proper confinement of the NSs. On the other hand, the structures with H equal to 0.4 or 0.6 times the NSs diameter promote the formation of confined and ordered monolayers at 2000 rpm and 2500 rpm.

So far, the selection of the optimal self-assembly parameters has been based on a qualitative analysis of the SEM images. To establish the efficiency of the DSA of NSs inside the hexagonal templates in a more rigorous way, the ordering process should be assessed quantitatively. To this goal, the SEM micrographs were processed with a user-defined image-processing routine based on Delaunay triangulation, measuring the orientation of HCP domains. The software recognises the domains and classify them according to their rotations in the angular range between −30° and 30°. The analysis was conducted on the hexagonal templates with H/D ratios of 0.4 and 0.6 highlighted in green in Figure 3. The results are collected in Figure 4, and report the normalised distributions of the orientation of the confined monolayer in different geometrical and dynamic conditions. Such angular distributions are centred on 0° indicating an alignment to the templates edges, while slight deviations in the orientation broaden the distributions. These can be accounted for by calculating the integral of the curve which gives the percent occurrence of the domains in a given orientation range. In the hexagonal templates with H/D ratio of 0.4 and 0.6, spinning speed of 2000 rpm lead to 43.5% and 56.1% of domains with orientation comprised between −10° and 10°, as shown in Figure 4a,b, respectively. In the graphs in Figure 4c,d, the percentage of aligned domains increases to 54.8% and 68.3% when the spin coating speed is set to 2500 rpm for H = 0.4·D and H = 0.6·D, respectively. This quantitative result clearly highlights that templates with H = 0.6·D induce a better ordering of the NSs when deposited at high spinning speed.

The confinement process was tested also inside linear templates, chosen for its simple realisation by DLW lithography, using the same optimal spinning conditions. The SEM micrographs reported in Figure 5a–d show the outcome of the SA process in the linear templates. Similarly to what was observed for the hexagonal templates, the micrographs coloured in orange (Figure 5a) and red (Figure 5d) correspond to unsuitable conditions for the DSA. Conversely, the templates with H/D ratio of 0.4 and 0.6 promote the formation of a confined self-assembled monolayer, as shown in Figure 5b,c. Also, in this case, the SEM micrographs were processed by Delaunay triangulation to evaluate the ordering process in terms of the domains orientation. The results of this study, reported in the graphs in Figure 5e,f, outline an angular distributions centred on 0° with narrow peaks including 89% of domains in the range from −10° to 10° for H = 0.4·D. This percentage rise up to 96.2% inside structures with H = 0.6·D.

According to this result, the linear templates induce a finer orientation constraint than the hexagonal ones, as they presented a regular shape and uniform width along their length as visible in Figure 5. On the other hand, the hexagonal structures presented some rounded features that may constitute a cause for the lower quality of the ordering process. Moreover, the dimension of the templates may differ from the pattern design causing incommensurability and the generation of defects in the colloidal lattice.

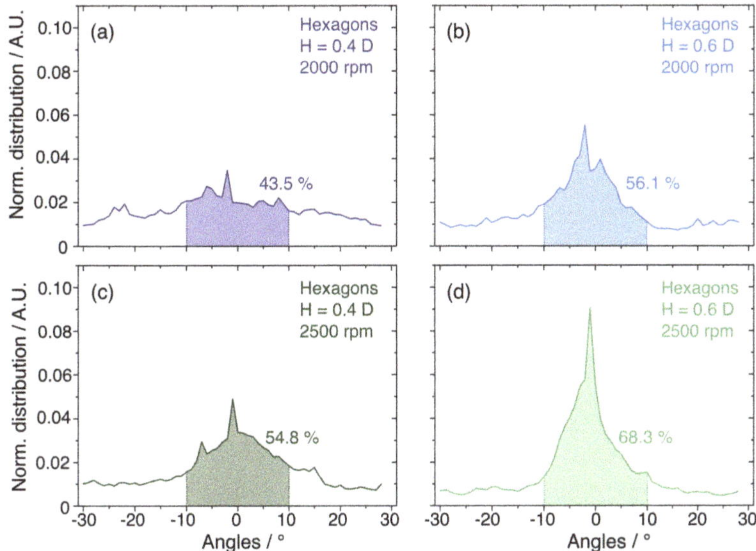

Figure 4. Line graphs reporting the normalised distribution of the domains orientation in the NSs monolayer confined in hexagonal graphoepitaxy structures with (**a**) H = 0.4·D at 2000 rpm, (**b**) H = 0.6·D at 2000 rpm, (**c**) H = 0.4·D at 2500 rpm and (**d**) H = 0.6·D at 2500 rpm. The orientation is evaluated in the angular range between −30° and 30°. The percentage of domains aligned to the templates edges is calculated through the integral of the curve in the range from −10° to 10° and the area and percent value are reported for each curve.

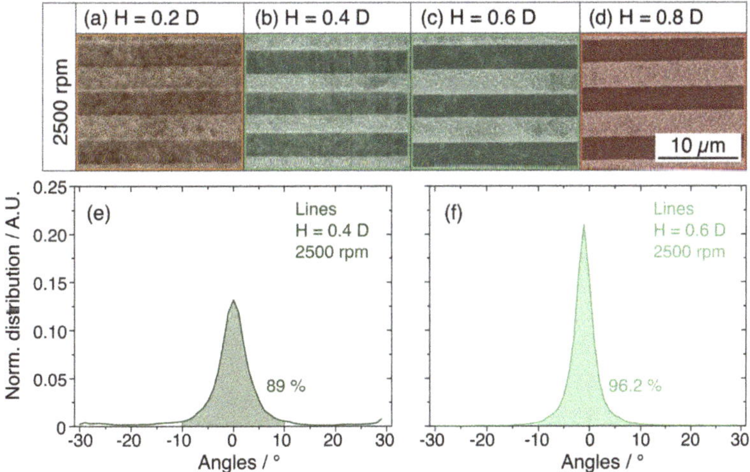

Figure 5. (**a–d**) SEM images of the NSs assemblies inside linear templates. The height of the constraining walls is varied from 0.2 to 0.8 times the diameter of the NSs across the images. (**e,f**) Line graphs reporting the normalised distribution of the domains orientation. The highlighted area corresponds to the percentage of domains aligned to the templates sidewalls within ±10°.

3.2. Nanospheres Confinement

Although the optimisation of the geometry and process parameters have led to a good result in terms of NSs ordering within the templates, the confinement process can be further investigated by considering the defectivity at the edge of the templates and the presence of residual nanospheres on the mesas, observed in Figures 3 and 5. We performed an AFM analysis of the confinement process focusing our attention on the height profile of the confined nanospheres in the two studied morphologies (i.e., hexagonal and linear) with H = 0.6·D.

Figure 6a reports an AFM map acquired on the hexagonal template. The height profile in Figure 6b indicates that, when good confinement is achieved, the NSs are perfectly aligned inside the hexagonal structure and exceed the mesa by $\Delta_{conf} = (96 \pm 4)$ nm. This value is quite similar to the one expected for H = 0.6·D as the difference between the sidewalls height and the sphere diameter. The AFM maps acquired in proximity of a defect (Figure 6c) and the corresponding height profile (Figure 6d) show an irregular arrangement of NSs. The nanosphere #2, closest to the confining wall, is found at the level $\Delta_{hex} = (167 \pm 1)$ nm above the mesa structure, whereas the NSs #3 and #4 are correctly confined at the level $\Delta_{conf} = (90 \pm 5)$ nm.

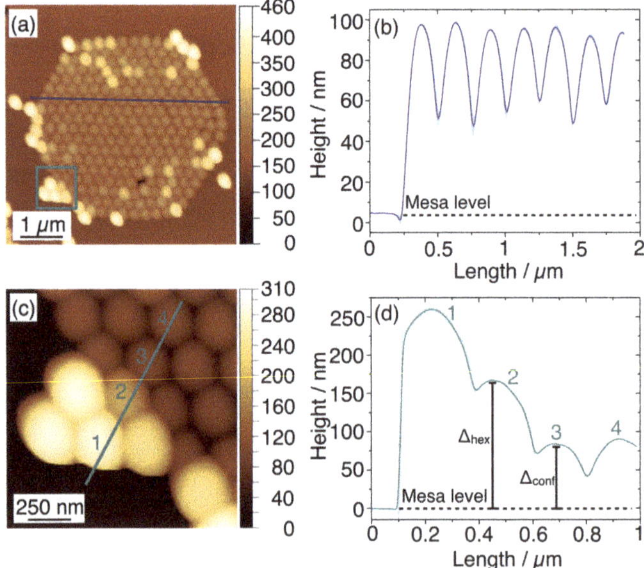

Figure 6. (**a**) AFM micrograph acquired on a hexagonal confining structure filled with a monolayer of NSs. The blue line marks the height profile in the line graph (**b**), where the shadowed region indicates the uncertainty of three repeated measurements. (**c**) The second AFM topographic image is acquired in the framed area in (**a**) and the line profile and numbered nanospheres are indicated. (**d**) The corresponding height profile reports nanosphere #2 is found at a higher level with respect from the mesa, $\Delta_{hex} = (167 \pm 1)$ nm, with respect to NSs #3 and #4.

Figure 7a,b reports the AFM micrograph acquired on a linear template and the corresponding height profile, respectively. When the NSs are well confined inside the template (e.g., the NSs labelled as #4 and #5), they lay at the same level for which $\Delta_{conf} = (86 \pm 2)$ nm. By approaching the side walls, the height of the nanospheres increases and NS #3 is separated from the top of the mesa by $\Delta_{lin} = (136 \pm 2)$ nm.

From this analysis, we observed the top of well-confined nanospheres to be at the level Δ_{conf} from the mesa, approximately equal to the difference between the diameter D and the sidewalls height H. When the separation exceeded this quantity, such as for Δ_{hex} and Δ_{lin} larger than Δ_{conf}, we observed the onset of a defect and the accumulation of unconfined NSs on the mesas. Given that Δ_{lin} was lower than the corresponding Δ_{hex}, the linear templates offered a better confinement of the nanospheres with respect to the hexagonal structures. In both templates, the observed distortions from the HCP symmetry can be due to several reasons, including local defectivity in the lithographic template, incommensurability of the graphoepitaxy structures or polydispersity of the nanospheres. These defects can be largely reduced by improving the combination of DLW lithography and RIE to obtain high regularity of the templates and fidelity to the pattern design. A possible strategy to limit the accumulation of excess NSs could be to graft hydrophobic polymer chains on the surface of the mesa.

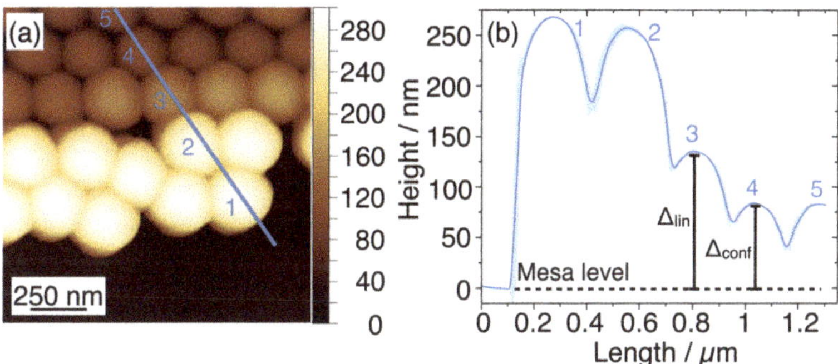

Figure 7. (**a**) AFM micrograph acquired on the NSs monolayer inside a linear template. The coloured line indicates the height profile evaluated on the topographic image. (**b**) The corresponding line graph presents the height variation as a function of lateral displacement. The value of the distance between NSs #3 and the top of the mesa is reported as $\Delta_{\text{lin}} = (136 \pm 2)$ nm.

Despite some local defectivity, the use of DLW lithography and RIE makes it simple to tailor the templates with H/D ratio fixed at 0.6 to confine NSs with different dimensions, as shown in Figure 8a,c for NSs with a diameter of 200 nm and 400 nm, respectively.

One common application of NSL consists in the realisation of triangular metallic nanoparticles as substrates for surface-enhanced Raman spectroscopy (SERS) applications, for the possibility to tune their geometrical features to match different excitation wavelengths [26]. DSA-NSL constitute a versatile solution to improve the uniformity and reproducibility in the fabrication of such substrates to benefit their spectroscopic responses, as it can be employed in the production of these and other metallic arrays with regular orientation and a high degree of order, as shown for example in Figure 8b,d.

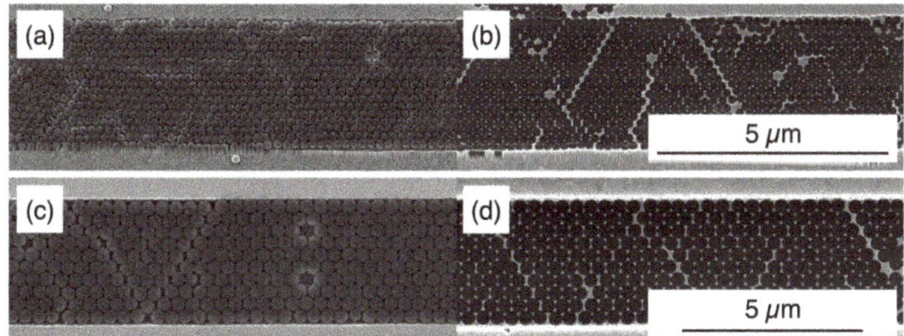

Figure 8. NSs with different sizes—(**a**) 200 nm and (**c**) 400 nm—are confined in a monolayer inside linear templates where the ratio H/D is kept constant at 0.6. This lithographic mask can be used for the realisation of arrays of gold nanotriangles (**b,d**) with tunable dimensions to match the excitation wavelengths for the sensing of different SERS-active analyte species.

4. Conclusions

In this work, we investigated the confinement and ordering process of the self-assembling NSs by changing the deposition parameters and the height of the confining walls in templates with two different shapes. The most appropriate conditions for the DSA-NSL where highlighted by a systematic SEM analysis correlated by the evaluation of the HCP orientation by image processing and atomic force microscopy measurements. High spinning speed of 2500 rpm were found to be necessary to let the NSs overcome the physical barriers of the templates. Sidewalls height H was found to provide proper confinement conditions at 0.6 times the NSs diameter.

DSA-NSL inside linear templates, with the previously stated geometrical and dynamic conditions, resulted in a confined monolayer aligned to the template for 96.2%. The knowledge on the DSA process and the control over the geometry through DLW lithography and RIE, allow to direct the SA of colloidal NSs to obtain single-grain crystals with uniform orientation and regular shape over large area. The optimised fabrication protocol could extend the versatility of DSA-NSL for applications requiring different geometries. The linear structures, for example, can be employed to confine the nanostructures in microfluidic channels for multiplexed analysis [27]. Moreover, hexagonal and circular structures with micrometric sizes can serve in site-specific incubation for different analytes in sensing applications, where the templates are easily recognised by optical microscopy to find the area of analysis.

Author Contributions: The authors individual contributions are the following: conceptualisation, E.C. and F.F.L.; methodology, E.C., F.F.L., M.F., M.T. and K.S.; software, E.C.; formal analysis, E.C., F.F.L. and M.T.; writing–original draft preparation, E.C., F.F.L., M.F., M.T., K.S., N.D.L., R.G. and L.B.; supervision, F.F.L., L.B. and N.D.; project administration, L.B.; funding acquisition, N.D.L. and L.B. All authors have read and agreed to the published version of the manuscript.

Funding: The project 16ENV07 Aeromet has received funding from the EMPIR programme co-financed by the Participating States and from the European Union's Horizon 2020 research and innovation programme.

Conflicts of Interest: The authors declare no conflicts of interest. The funders had no role in the design of the study; in the collection, analyses, or interpretation of data; in the writing of the manuscript; or in the decision to publish the results.

References

1. Colson, P.; Henrist, C.; Cloots, R. Nanosphere lithography: A powerful method for the controlled manufacturing of nanomaterials. *J. Nanomater.* **2013**, *2013*, 21. [CrossRef]
2. Morandi, V.; Marabelli, F.; Amendola, V.; Meneghetti, M.; Comoretto, D. Colloidal photonic crystals doped with gold nanoparticles: spectroscopy and optical switching properties. *Adv. Funct. Mater.* **2007**, *17*, 2779–2786. [CrossRef]
3. Li, N.; Pang, S.; Yan, F.; Chen, L.; Jin, D.; Xiang, W.; Zhang, D.; Zeng, B. Window-assisted nanosphere lithography for vacuum micro-nano-electronics. *AIP Adv.* **2015**, *5*, 047101. [CrossRef]
4. Toma, M.; Loget, G.; Corn, R.M. Fabrication of broadband antireflective plasmonic gold nanocone arrays on flexible polymer films. *Nano Lett.* **2013**, *13*, 6164–6169. [CrossRef] [PubMed]
5. Liu, C.C.; Franke, E.; Mignot, Y.; Xie, R.; Yeung, C.W.; Zhang, J.; Chi, C.; Zhang, C.; Farrell, R.; Lai, K. Directed self-assembly of block copolymers for 7 nanometre FinFET technology and beyond. *Nat. Electron.* **2018**, *1*, 562. [CrossRef]
6. Stoykovich, M.P.; Müller, M.; Kim, S.O.; Solak, H.H.; Edwards, E.W.; De Pablo, J.J.; Nealey, P.F. Directed assembly of block copolymer blends into nonregular device-oriented structures. *Science* **2005**, *308*, 1442–1446. [CrossRef]
7. Wan, L.; Ruiz, R.; Gao, H.; Patel, K.C.; Albrecht, T.R.; Yin, J.; Kim, J.; Cao, Y.; Lin, G. The limits of lamellae-forming PS-b-PMMA block copolymers for lithography. *ACS Nano* **2015**, *9*, 7506–7514. [CrossRef]
8. Liu, G.; Thomas, C.S.; Craig, G.S.; Nealey, P.F. Integration of Density Multiplication in the Formation of Device-Oriented Structures by Directed Assembly of Block Copolymer–Homopolymer Blends. *Adv. Funct. Mater.* **2010**, *20*, 1251–1257. [CrossRef]
9. Ferrarese Lupi, F.; Giammaria, T.J.; Miti, A.; Zuccheri, G.; Carignano, S.; Sparnacci, K.; Seguini, G.; De Leo, N.; Boarino, L.; Perego, M. Hierarchical Order in Dewetted Block Copolymer Thin Films on Chemically Patterned Surfaces. *ACS Nano* **2018**, *12*, 7076–7085. [CrossRef]
10. Perego, M.; Andreozzi, A.; Vellei, A.; Lupi, F.F.; Seguini, G. Collective behavior of block copolymer thin films within periodic topographical structures. *Nanotechnology* **2013**, *24*, 245301. [CrossRef]

11. Grzelczak, M.; Vermant, J.; Furst, E.M.; Liz-Marzan, L.M. Directed self-assembly of nanoparticles. *ACS Nano* **2010**, *4*, 3591–3605. [CrossRef] [PubMed]

12. Yin, Y.; Lu, Y.; Gates, B.; Xia, Y. Template-assisted self-assembly: A practical route to complex aggregates of monodispersed colloids with well-defined sizes, shapes, and structures. *J. Am. Chem. Soc.* **2001**, *123*, 8718–8729. [CrossRef] [PubMed]

13. Rycenga, M.; Camargo, P.H.; Xia, Y. Template-assisted self-assembly: A versatile approach to complex micro-and nanostructures. *Soft Matter* **2009**, *5*, 1129–1136. [CrossRef]

14. Xia, Y.; Yin, Y.; Lu, Y.; McLellan, J. Template-assisted self-assembly of spherical colloids into complex and controllable structures. *Adv. Funct. Mater.* **2003**, *13*, 907–918. [CrossRef]

15. Varghese, B.; Cheong, F.C.; Sindhu, S.; Yu, T.; Lim, C.T.; Valiyaveettil, S.; Sow, C.H. Size selective assembly of colloidal particles on a template by directed self-assembly technique. *Langmuir* **2006**, *22*, 8248–8252. [CrossRef]

16. Cara, E.; Mandrile, L.; Lupi, F.F.; Giovannozzi, A.M.; Dialameh, M.; Portesi, C.; Sparnacci, K.; De Leo, N.; Rossi, A.M.; Boarino, L. Influence of the long-range ordering of gold-coated Si nanowires on SERS. *Sci. Rep.* **2018**, *8*, 11305. [CrossRef]

17. Chee, K.W.; Guo, W.; Wang, J.R.; Wang, Y.; Chen, Y.e.; Ye, J. Tuning photonic crystal fabrication by nanosphere lithography and surface treatment of AlGaN-based ultraviolet light-emitting diodes. *Mater. Des.* **2018**, *160*, 661–670. [CrossRef]

18. Horrer, A.; Schäfer, C.; Broch, K.; Gollmer, D.A.; Rogalski, J.; Fulmes, J.; Zhang, D.; Meixner, A.J.; Schreiber, F.; Kern, D.P.; et al. Nanosphere Lithography: Parallel Fabrication of Plasmonic Nanocone Sensing Arrays (Small 23/2013). *Small* **2013**, *9*, 4088. [CrossRef]

19. Kara, S.; Keffous, A.; Giovannozzi, A.; Rossi, A.; Cara, E.; D'Ortenzi, L.; Sparnacci, K.; Boarino, L.; Gabouze, N.; Soukane, S. Fabrication of flexible silicon nanowires by self-assembled metal assisted chemical etching for surface enhanced Raman spectroscopy. *RSC Adv.* **2016**, *6*, 93649–93659. [CrossRef]

20. Krupinski, M.; Sobieszczyk, P.; Zieliński, P.; Marszałek, M. Magnetic reversal in perpendicularly magnetized antidot arrays with intrinsic and extrinsic defects. *Sci. Rep.* **2019**, *9*, 13276. [CrossRef]

21. Yao, Y.; Zhang, L.; Orgiu, E.; Samorì, P. Unconventional Nanofabrication for Supramolecular Electronics. *Adv. Mater.* **2019**, *31*, 1900599. [CrossRef] [PubMed]

22. Brassat, K.; Assion, F.; Hilleringmann, U.; Lindner, J.K. Self-organization of nanospheres in trenches on silicon surfaces. *Phys. Status Solidi (A)* **2013**, *210*, 1485–1489. [CrossRef]

23. Parchine, M.; McGrath, J.; Bardosova, M.; Pemble, M.E. Large area 2D and 3D colloidal photonic crystals fabricated by a roll-to-roll Langmuir–Blodgett method. *Langmuir* **2016**, *32*, 5862–5869. [CrossRef]

24. Denkov, N.; Velev, O.; Kralchevsky, P.; Ivanov, I.; Yoshimura, H.; Nagayama, K. Two-dimensional crystallization. *Nature* **1993**, *361*, 26. [CrossRef]

25. Kralchevsky, P.A.; Denkov, N.D. Capillary forces and structuring in layers of colloid particles. *Curr. Opin. Colloid Interface Sci.* **2001**, *6*, 383–401. [CrossRef]

26. Haes, A.J.; Haynes, C.L.; McFarland, A.D.; Schatz, G.C.; Van Duyne, R.P.; Zou, S. Plasmonic materials for surface-enhanced sensing and spectroscopy. *MRS Bull.* **2005**, *30*, 368–375. [CrossRef]

27. Strehle, K.R.; Cialla, D.; Rösch, P.; Henkel, T.; Köhler, M.; Popp, J. A reproducible surface-enhanced Raman spectroscopy approach. Online SERS measurements in a segmented microfluidic system. *Anal. Chem.* **2007**, *79*, 1542–1547. [CrossRef]

nanomaterials

Review

Recent Advances in Sequential Infiltration Synthesis (SIS) of Block Copolymers (BCPs)

Eleonora Cara [1,*], Irdi Murataj [1,2], Gianluca Milano [1], Natascia De Leo [1], Luca Boarino [1] and Federico Ferrarese Lupi [1,*]

[1]   Advanced Materials and Life Sciences, Istituto Nazionale di Ricerca Metrologica (INRiM),
      Strada delle Cacce 91, 10135 Turin, Italy; irdi.murataj@polito.it (I.M.); g.milano@inrim.it (G.M.);
      n.deleo@inrim.it (N.D.L.); l.boarino@inrim.it (L.B.)
[2]   Dipartimento di Scienza Applicata e Tecnologia, Politecnico di Torino, Corso Duca degli Abruzzi 24,
      10129 Turin, Italy
*   Correspondence: e.cara@inrim.it (E.C.); f.ferrareselupi@inrim.it (F.F.L.)

Abstract: In the continuous downscaling of device features, the microelectronics industry is facing the intrinsic limits of conventional lithographic techniques. The development of new synthetic approaches for large-scale nanopatterned materials with enhanced performances is therefore required in the pursuit of the fabrication of next-generation devices. Self-assembled materials as block copolymers (BCPs) provide great control on the definition of nanopatterns, promising to be ideal candidates as templates for the selective incorporation of a variety of inorganic materials when combined with sequential infiltration synthesis (SIS). In this review, we report the latest advances in nanostructured inorganic materials synthesized by infiltration of self-assembled BCPs. We report a comprehensive description of the chemical and physical characterization techniques used for *in situ* studies of the process mechanism and *ex situ* measurements of the resulting properties of infiltrated polymers. Finally, emerging optical and electrical properties of such materials are discussed.

Keywords: sequential infiltration synthesis; block copolymer; self-assembly

Citation: Cara, E.; Murataj, I.; Milano, G.; De Leo, N.; Boarino, L.; Ferrarese Lupi, F. Recent Advances in Sequential Infiltration Synthesis (SIS) of Block Copolymers (BCPs). *Nanomaterials* **2021**, *11*, 994. https://doi.org/10.3390/nano11040994

Academic Editor: Sebastien Lecommandoux

Received: 8 March 2021
Accepted: 4 April 2021
Published: 13 April 2021

Publisher's Note: MDPI stays neutral with regard to jurisdictional claims in published maps and institutional affiliations.

1. Introduction

The seek for novel materials with tailored properties has been of great interest among the scientific community over the last decades. The ability to fabricate nanostructured inorganic materials with high degree of control on morphology and dimensions, led to advanced materials with boosted performances in different research fields, such as nano-lithography [1–4], photonics [5], biomedicine [6,7] and energy [8,9]. The realization of wide-area periodic nanopatterns is currently the subject of many efforts by the micro-electronics industry, pushing the development of next-generation electronic and optical devices. At the moment, conventional lithographic techniques (i.e., optical and electron lithographies) represent the workhorse of micro and nanoscale manufacturing. Over the last years, their technological improvements determined significant advances, approaching the fundamental requirements demanded by the continuous downscale of device features. However, conventional lithographic techniques are now facing their intrinsic technological and economic limits [10] in terms of large-scale pattern definition and material deposition.

Among alternative nanopatterning methods, self-assembled materials such as block copolymers (BCPs) demonstrated to be very valuable in the pursuit of the shrinkage of electronic and optical devices, offering large scale scalability and a ready integration in the manufacturing processes [10,11]. The self-assembly of BCPs, in particular, represents a cost-effective bottom–up approach with high throughput, able to provide highly dense periodic patterns at the nanoscale in the typical range of 10–100 nm. Compared to optical and electron lithography, the self-assembly of BCPs relies on the in-parallel self-registration of amphiphilic BCPs, driven by the chemical incompatibility between the constituent

25

blocks. A high degree of control on self-assembled nanostructures, in terms of orientation [12,13], long-range ordering [14–16], morphology [17] and feature size [18,19] is related to the ability to finely tune the substrate functionalization, annealing conditions and the characteristic parameters of BCPs (i.e., molecular weight and composition). The potential use of BCPs for several semiconductor industry technologies was recently assessed by Liu et al. [10]. By a direct comparison of directed self-assembly (DSA) of BCPs with conventional multi-step patterning approaches, such as self-aligned double/quadruple patterning (SADP/SAQP); the authors demonstrated the feasibility of applying BCP nanopatterning in the fabrication of 7 nm node fin field-effect transistors (FinFETs) in high-volume manufacturing testing. In addition, the pattern quality of fabricated patterns, in terms of critical dimension and pitch uniformity, was reported to be sufficient for integrated circuit layer manufacturing. The overall lower processing cost and high scalability provided by self-assembly of BCPs could also pave the way for the fabrication of self-assembled crossbar arrays of memristive devices for the realization of next-generation computing architectures, as also underlined in the roadmap on emerging hardware and technology for machine learning [20]. The great flexibility provided by the BCPs offers the opportunity to employ them as a nanopatterning tool for the design and fabrication of a wide range of functional materials. In particular, when combined with emerging synthetic routes as sequential infiltration synthesis (SIS), BCPs represent ideal templates for the synthesis of hybrid organic/inorganic or all-inorganic nanostructured materials with potential applications spanning from nanoelectronics [21] to photonics [22] and optics [23]. The SIS process is a vapor-phase and solvent-free process based on atomic layer deposition (ALD), generally used for the inclusion of inorganic materials into polymer templates. SIS consists of the cyclic exposure of polymers to a vapor-phase metal–organic precursor and an oxidizing agent (H_2O, H_2O_2, O_3), which leads to the formation of organic/inorganic hybrid materials. When SIS is applied to self-assembled BCPs, the metal–organic precursors are selectively entrapped inside the polar homopolymer composing the BCPs. Subsequent removal of the polymeric species, obtained whether by polymer ashing [24] or plasma etching [25], reveals a nanostructured metal oxide whose morphology perfectly replicates that of the BCPs template [26], as schematized in Figure 1.

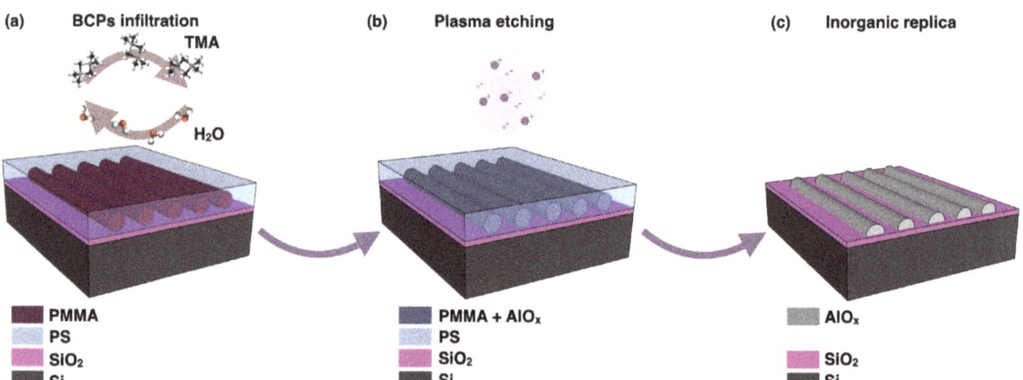

Figure 1. Schematic process flow of the sequential infiltration synthesis of block copolymers (BCPs). (**a**) ALD cycles with gaseous precursors (for instance trimethyl aluminum (TMA) and water). (**b**) Removal of the uninfiltrated polymeric component by plasma etching. (**c**) Inorganic replica of the BCPs template.

Although sharing the same equipment and metal–organic precursors, the processing parameters of SIS substantially differ from that of conventional ALD processes, widely used for the conformal deposition of inorganic thin films on solid substrates (Figure 2a). Indeed, in conventional ALD, the cyclic exposures to the metal–organic precursors are typically very short, at low partial pressure and aimed at saturating all the reactive sites on the substrate surface in a self-limiting fashion. By contrast, in SIS the goal is to dissolve, diffuse and entrap the precursors throughout the entire BCPs film thickness (Figure 2b), thus requiring higher exposure partial pressures and times [27–29]. The extensive research over the last years has referred to SIS with different terminologies i.e., vapor phase infiltration (VPI) [30], micro-dose infiltration synthesis (MDIS) [31] and multipulse vapor infiltration (MPI) [32]. Although each process indicates a different precursor dosing sequence, they all rely on the same fundamental phenomenology [30].

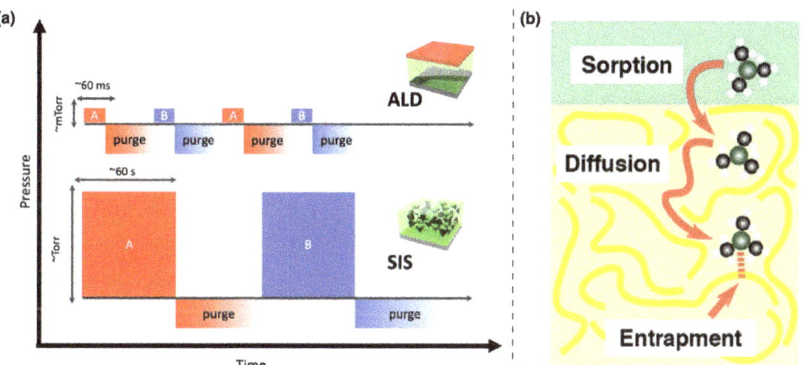

Figure 2. (**a**) Schematic comparison of conventional atomic layer deposition (ALD) and sequential infiltration synthesis (SIS) protocols. Reproduced and adapted from reference [27]. Copyright 2019, AIP Publishing. (**b**) Schematic illustration of metal–organic precursor infiltration process into polymers. Adapted with permission from reference [29]. Copyright 2019 American Chemical Society.

Here, we report recent advances and perspectives of the SIS process, with a specific focus on the synthesis of nanostructured materials by BCPs templates. Great attention is dedicated to the discussion of *in situ* and *ex situ* spectroscopic and microscopic characterization techniques adopted for an exhaustive comprehension of the process mechanism and morphological, compositional and structural characterization. Subsequently, in this review, we address the emerging optical and electrical properties of infiltrated materials with potential technological impact on the development of novel devices.

2. SIS Processing and Mechanism

The SIS of BCPs follows a Lewis acid–base interaction between the metal–organic precursors (Lewis acids) and functional groups of the polar domains (Lewis bases). Being polystyrene *block* poly(methyl methacrylate) (PS-*b*-PMMA) the prototypical BCPs, widely used as reference material for the study of the self-assembly process, a lot of effort has been dedicated to the understanding of the mechanism involved in the SIS [25,33,34]. Early studies on the synthesis of aluminum oxide (AlO_x) obtained after the cyclic exposure of PS-*b*-PMMA to trimethylaluminum (TMA) and water, demonstrated that the TMA–PMMA interaction follows a two-step adsorption [35]. The first step consists in the formation of a Lewis adduct obtained by the reversible coordination of TMA to the carbonyl (C=O) of the ester groups of PMMA, then followed by a slow conversion into covalent Al–O bond [36]. Subsequent exposure to water determines the formation of O–Al–OH species, due to the oxidation of bonded TMA, that act as nucleation and growth sites for AlO_x in the following SIS cycles [31]. The lack of polar functional groups in PS implies the absence of any interaction of the precursors with the aforementioned homopolymer. Consequently, PS acts as a diffusive channel for the transport of the precursors to the reactive sites of PMMA [33]. A similar behavior is also found in the statistical copolymer polystyrene-*stat*-poly(methyl methacrylate) (PS-*stat*-PMMA). However, the TMA diffusivity is affected as the MMA unit content in the polymer film varies, reaching a maximum value for MMA fraction of 0.56 [37]. The inert properties of PS towards metal–organic precursors has been recently exploited for the uniform coating of freestanding nanoparticles. By applying the SIS on resting nanoparticles on a PS layer, the precursors can diffuse through the underlying PS and reach the reactive sites on the bottom part of the nanoparticles. This allows the growth of the metal oxide on nanoparticles even on the side in contact with the substrate, otherwise not possible with standard ALD process [38].

2.1. Polymer Selectivity

The search for a comprehensive insight into the SIS mechanism has also been extended to polymers with amides and carboxylic acids functional groups, such as poly(vinylpyrrolidone) (PVP) and poly(acrylic acid) (PAA), respectively. While PVP shows similar reactivity to PMMA, forming a reversible Lewis adduct $C=O\cdots Al--(CH_3)_3$, in PAA the presence of an acidic proton determines the direct covalent Al–O bonding through a pericyclic reaction [39] (Figure 3).

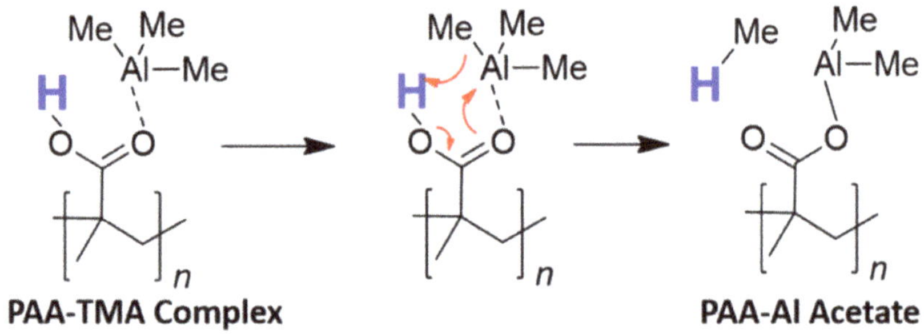

Figure 3. Proposed pericyclic mechanism for trimethylaluminum (TMA) and poly(acrylic acid) (PAA) reaction. Adapted with permission from reference [39]. Copyright 2019 American Chemical Society.

Different polymers with carbonyl-containing functional groups, therefore, show substantial differences in the interaction dynamics with the metal–organic precursors. Biswas et al. [40] recently reported that, although sharing the same ester functional groups, poly(ϵ-caprolactone) (PCL) interacts more strongly with TMA and $TiCl_4$ compared to PMMA, showing nearly total saturation of the available $C=O$ sites for both precursors. The higher reactivity of PCL is to be found in the polymer backbone positioning of the carbonyl groups that confers a higher nucleophilicity compared to the side chain $C=O$ groups in PMMA, resulting in a stronger Lewis acid–base interaction with metal–organic precursors.

The increasing research on new polymers with oxygen-containing functional groups pushes forward the achievement of direct selective growth of different nanostructured metal oxides as ZnO, TiO_x and VO_x that otherwise would require pre-infiltration of AlO_x [41–43]. As an example, Yi et al. [44] reported how cyclic ether groups of polystyrene-*block*-poly(epoxyisoprene) (PS-*b*-PIO) act as effective templates for the direct infiltration synthesis with TMA, diethylzinc (DEZ), titanium isopropoxide ($Ti(OiPr)_4$) and vanadyl isopropoxide ($VO(OiPr)_3$) thanks to a greater Lewis basicity of cyclic ether groups when compared to the ester group of PMMA.

Surprisingly, the same authors found a selective growth of ZnO and AlO_x in polyisoprene domains of polystyrene-*block*-poly(1,4-isoprene) (PS-*b*-PI) BCPs even though lacking any polar ligand group, suggesting that the Lewis acid–base interaction alone is insufficient to fully describe the precursor entrapment. A first attempt of explanation on how alkene functional groups in PS-*b*-PI can play a role in entrapping metal–organic precursors was given by attributing it to the high permeability of PI to a given precursor. Lately, a more in-depth assessment of the mechanism involving SIS with DEZ in cis-polyisoprene, revealed that pre-heating treatments play a key role in increasing the load of metal–organic precursors by inducing chemical changes to cis-polyisoprene. Indeed, pre-heated films undergo partial oxidation, which introduces new $C=O$ functional groups responsible for the increased metal–organic entrapment [45]. A list of relevant references focusing on SIS on different polymers and functional groups is presented in Table 1.

Table 1. Polymers sorted by functional groups, utilized as templates for sequential infiltration synthesis (SIS) in the recent literature and the corresponding references.

Functional Groups	Polymers	References
Alkenes	PS-*b*-PI	[44,45]
Amides	PVP	[39]
Carboxylic acids	PAA	[39]
Esters	PS-*b*-PMMA PCL	[25,33–36] [40]
Epoxydes	PS-*b*-PIO	[44]
Pyridines	PS-*b*-P2VP PS-*b*-P4VP	[31,40] [46,47]

The extensive literature on SIS of nanostructured metals and metal oxides as AlO_x [33], SiO_x, TiO_x [48], ZnO [31], W [25] and WO_{3-x} [42] proved self-assembled BCPs templates as a promising tool for nanopatterning applications, thus pushing the research to the development of SIS for new semiconducting materials such as In_2O_3, Ga_2O_3 [49] and SnO_x [43].

2.2. Diffusion

When comparing the phenomena involved in SIS (i.e., sorption, diffusion and entrapment) to ALD, a higher complexity is determined by the larger number of experimental design parameters that need to be taken into account, namely: temperature, pressure, pre-treatments, precursor and oxidizing agent exposure times, purge time and polymer–precursor interaction [30]. The ability to perform deep infiltration of inorganic materials into polymers represents one of the fundamental aspects to expand the technological impact of SIS on a wide range of applications. The diffusion of inorganic precursors into polymeric templates, although being of critical importance, suffers from limitations in terms of depth of penetration that affect the inorganic material mass incorporation and pattern quality [43]. Different strategies have been recently developed in order to increase the effective diffusion of metal–organic precursors into polymer templates. Examples of infiltration of PS-*b*-P4VP (polystyrene-*block*-poly(4-vinylpyridine)) in polar swelling solvents (i.e., ethanol), show a more efficient infiltration thanks to the introduction of additional porosity channels [46]. The swelling-assisted SIS is a method based on the immersion of BCP films into a polar solvent prior to the infiltration. The incorporation of polar organic solvents into the polar domains of the BCP, upon subsequent drying, determines the formation of interconnected pores in the typical range of 10–50 nm. These pores act as effective pathways for the delivery of the metal–organic precursors throughout the BCP film thickness [50]. Thus, they enable the access of the metal–organic precursors to all the available sites. This results in a two-fold increase of the amount of synthesized AlO_x, proving to be a valid approach also for the synthesis of porous multicomponent heterostructures [47]. Higher amounts of precursor molecules available for a more efficient diffusion into the polymer, can be delivered by modifying the conventional SIS process parameters. MDIS is a modified infiltration synthesis protocol which consists in repeating the precursor dosing multiple times while still maintaining static vacuum. The higher cumulative duration of precursor exposure in MDIS, when compared to conventional SIS protocol, determines a higher concentration of precursor molecules in the chamber. This translates into a higher number of molecules available to diffuse into the polymer, allowing the growth of a superior amount of material and a more uniform block-selective infiltration [31].

The control over precursors diffusion can also be exploited to expand the library of new multicomponent materials that can be synthesized with SIS. As recently reported by

Azoulay et al., by designing the diffusive time of TMA and DEZ into cylinder-forming PS-*b*-PMMA, they were able to simultaneously grow different metal oxides at designated locations. Short TMA exposure times determined a shallow infiltration of the PMMA cylinder domains, whereas longer exposures of DEZ allowed a deeper diffusion into the entire film depth leading to the synthesis of an inorganic nanorod array of AlO_x−ZnO heterostructures [51]. The full comprehension of the synthetic process requires also to consider the polymer–precursor interaction and its relation to the temperature, since their significant influence on the precursor effective diffusion. A clear insight into the role of temperature on the SIS was given by Weisbord et al. in a recent publication [52]. In a temperature-dependent model, the authors predicted the existence of a balance point temperature of thermodynamic equilibrium ($\Delta G = 0$) for each polymer–precursor pair. At the balance point temperature, the forward and reverse polymer–precursor interactions satisfy the thermodynamic conditions for maximum mass gain (Figure 4a,b).

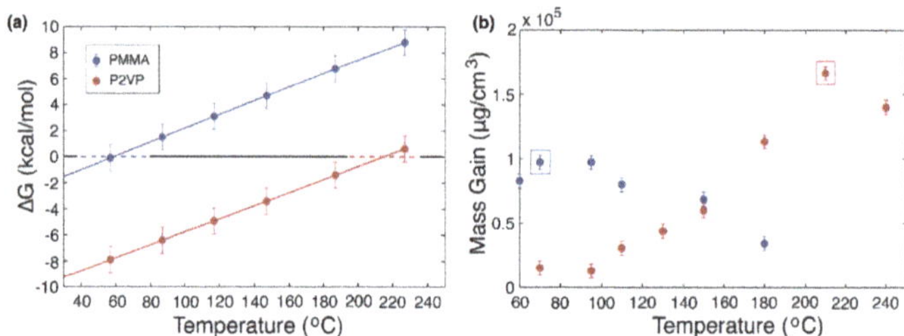

Figure 4. (**a**) Balance point temperature calculations for TMA-PMMA (poly(methyl methacrylate)) and TMA-P2VP (poly(2-vinylpyridine)) pairs and (**b**) relative experimental mass gain as a function of the temperature. Reproduced and adapted under the terms of Creative Commons Attribution 4.0 License from reference [52]. Copyright 2020 American Chemical Society.

The Lewis basicity of each polymer strongly influences the balance point temperature. For strong Lewis base polymers such as poly(2-vinylpyridine) (P2VP), high temperatures ($\approx$210 °C) are desired for maximum mass gain. However, at these temperatures self-assembled BCPs such as PS-*b*-P2VP cannot maintain their pattern and consequently undergo morphology rearrangement that prevents the pattern quality of the infiltrated material. To overcome this issue, a multi-temperature SIS process was proposed. By the combination of a first low-temperature (80 °C) SIS cycle followed by four SIS cycles at a higher temperature (150 °C) the authors were able to obtain a higher mass gain for PS-*b*-P2VP when compared to single-temperature processes. Although being far from the thermodynamic conditions of maximum mass gain, the mass of AlO_x accumulated in the first SIS cycle at (80 °C) prevents any BCP reconfiguration, preserving the vertically oriented cylinders pattern. Then, the subsequent high-temperature SIS cycles (150 °C) guarantee the highest mass growth (Figure 5).

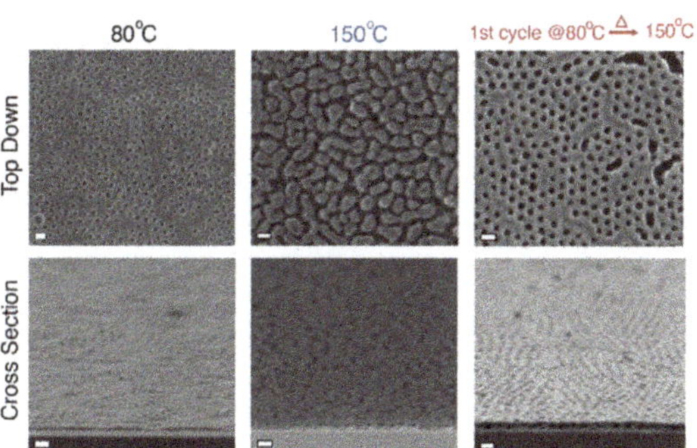

Figure 5. Top-down and cross-sectional scanning electron microscopy (SEM) images of AlO$_x$ nanopatterns obtained after SIS at 80 °C, 150 °C and multi-temperature processes. Scales bars are 100 nm. Reproduced and adapted under the terms of Creative Commons Attribution 4.0 License from reference [52]. Copyright 2020 American Chemical Society.

3. Characterization Techniques

The development of the SIS process in terms of fabrication has progressed rapidly in the latest years, implementing a wide choice of materials for precursors and polymers and a large set of varying parameters regulating the infiltration process. However, the complete comprehension of the process mechanism and the exhaustive characterization of the materials' properties have not yet followed through the expanding fabrication capabilities. Recent developments of lithographic, optical, mechanical and electrical applications of the SIS process require extensive characterization of the materials' properties. A large set of physical and chemical methods has been applied so far with the aim to characterize the infiltrated polymeric nanostructures. The interest of the SIS community has been pointed at both the chemistry of reactions involved among the gaseous precursors and the polymer and the reconstruction of the morphology of the oxides nanostructures from a compositional and dimensional point of view. *In situ* characterization techniques have been used to unravel the phenomenology of the infiltration process inside the ALD chamber, while *ex situ* methods have been dedicated to the characterization of the results of the process at the end of different number of ALD cycles conducted under the same conditions. Given the wide variety of precursors and polymeric species used in literature and different process parameters, specific results of the characterization vary from study to study. Hereafter, we discuss how the different characterization techniques have been adopted for the study of SIS and we highlight the major achievements in understanding the process.

3.1. Phenomenology of the Infiltration Process

In the latest years, several *in situ* methodologies has been used and adapted inside the infiltration process chamber to gain direct access to the steps of the precursors infiltration in the polymeric matrix, i.e., the sorption of the gas-phase precursor, the diffusion and the entrapment inside the polymer [53].

Fourier-transform infrared spectroscopy (FTIR) is a well-known spectroscopic method based on the monitoring of adsorption peaks at different vibration frequencies in the mid-infrared range, constituting a fingerprint spectrum and corresponding to the chemical interactions among the reactants involved in a process. Integrated into the ALD chamber, FTIR is used for the temporal evolution analysis of the reactions between the organometallic precursor and the polymer functional groups at different stages of the ALD process. Transmission and reflection FTIR allow identifying the relevant moieties and the specific bonds that are formed (positive peaks) or consumed (negative peaks) or shifted in the phases of the infiltration process when changing the reaction parameters [35]. The spectral features are subtracted by a reference spectrum, acquired on a pristine substrate [27].

A notable example of the information retrieved from such spectral analysis is found in references [35,36], where some early results on *in situ* transmission FTIR measurements on PMMA thin films infiltrated with TMA were presented. The authors hypothesized and verified that the TMA reaction with PMMA occurs in a two-step process. The TMA is quickly absorbed by carbonyl C=O and ester C–O–R moieties in PMMA, forming a weakly-bound intermediate complex that is then slowly converted into a covalent bond, generating Al–O [35]. The analysis of temperature, thickness and time-dependence of the adsorption gave a deeper understanding of the process kinetics. The FTIR study highlighted that the adsorption of TMA into the PMMA film is a diffusion-limited process requiring long exposures to reach saturation with a quadratic functional dependence to time. The same time-dependence was observed in the desorption of TMA during purge time with desorption 10 times longer than adsorption [36].

Recently, another work on *in situ* FTIR measurements extended the analysis to different combinations of precursors (i.e., TMA and $TiCl_4$) and polymers (i.e., PMMA, P2VP and PCL) to monitor the spectral changes of the reactive functional groups and kinetics of the adsorption and desorption processes [40]. Figure 6a,b report the absorption spectra of PCL acquired at the first and second SIS cycle at the two precursors' exposure steps. Spectrum 3a.1 revealed a complete loss of C=O feature upon TMA interaction with the polymer, a blue-shift of C–O–R peak corresponding to a modification of the bond length and the formation of a $AlCH_3$ complex. Upon the water dose, spectrum 3a.2 the C–O–R shift and aluminum complex peak are reversed indicating a loss of the surface species and complexed C–O–R. The C=O negative peak is not reversed indicating a unique irreversible covalent bond with TMA. Similar but less pronounced features are visible in spectrum 3b.1 corresponding to the first dose of $TiCl_4$ in PCL. The spectrum presents C=O and C–O–R negative features, consistent with their consumption and a positive peak corresponding to the formation of a C–Cl complex. In this case, a non-covalent complex formation can be deduced from the spectrum 3b.2, where the reversed C=O peak suggests the partial release of these groups interacting with Ti–Cl species. For both graphs, the second SIS cycle is characterized by the same features, only with reduced intensities. The histogram in Figure 6c summarizes the FTIR results for the analyzed homopolymers reporting the percentage consumption of the reactive functional groups at different steps of the first SIS cycle for the two used precursors. This graph highlights the strong and stable reactivity of PCL to both TMA and $TiCl_4$, allowing to identify PCL as a promising candidate for the infiltration process both as homopolymer and copolymer, matched with a non-reactive polymer such as PS.

Quartz crystal microbalance (QCM) gravimetry is quite often used in combination with *in situ* FTIR or alone to monitor the SIS process *in situ* [30,33,43,51,52,54]. It consists in employing a quartz crystal commonly used in deposition systems and modifying it with a thermally-equilibrated polymeric coating matching the polymer which is being infiltrated in the vacuum chamber of the ALD [54]. During the precursor adsorption and diffusion inside the polymer, the changes in the oscillation frequency of the quartz crystal are monitored and converted into the precursor mass uptake or loss of the polymer, through the knowledge of the material density and acoustic impedance. These features render QCM gravimetry a versatile technique, allowing to gain insights into the growth kinetics for every oxide in the SIS library [27,43,51] in both molecular layer deposition and etching processes [55].

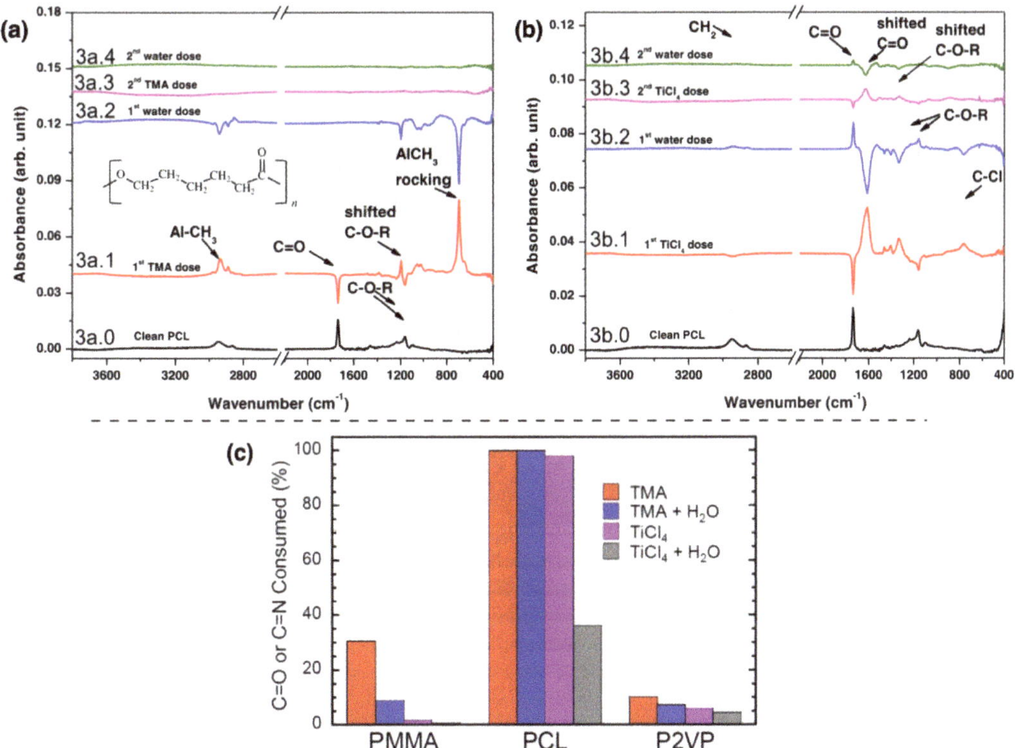

Figure 6. The adsorption spectra of poly(ϵ-caprolactone) (PCL) infiltrated with (**a**) TMA and (**b**) TiCl$_4$ are shown. The spectra from bottom to top are referred to the pristine polymer layer (black line), the first SIS cycle (red and blue lines) and the second SIS cycle (pink and green lines). The histogram in panel (**c**) summarizes the percentage consumption of C=O (for PMMA and PCL) and C=N (for P2VP) functional groups at different stages of the infiltration process. All panels are reproduced and adapted with permission from reference [40]. Copyright 2020 American Chemical Society.

The time-dependent measurements usually present an increase in the mass gain of the polymer during the exposure to the precursor, potentially reaching saturation with zero slope, followed by a mass loss in the purging step, when the unreacted reactants and byproducts are desorbed from the polymer. The slope of the mass gain in the different steps can provide information on the diffusivity of the precursors in the polymer. In Figure 7a, QCM gravimetric measurements are conducted on a PMMA thin film during the TiO$_2$ SIS process using TiCl$_4$ as precursor [33]. A large initial mass gain is displayed indicating a great diffusivity of the TiCl$_4$ precursor in the polymer, followed by a modest rate of mass uptake in the following steps. The slope of the desorption step provides information on the kinetics of the process. The steep mass loss during the exposure to H$_2$O vapor precursor in the TiO$_2$ infiltration of PMMA suggests a fast kinetics between water and the TiCl$_4$–PMMA complex and the release of different byproducts of such reaction [33]. Analogously, gravimetric measurements of the infiltration of two precursors, TMA for alumina and DEZ for zinc oxide growth, are reported in reference [51] for a self-assembled PS-*b*-PMMA film, revealing a much more abrupt and steep adsorption for TMA than for DEZ, thus indicating a faster diffusion for TMA. Gradual and long desorption of TMA from PMMA domains (not shown here) evidences a slow release of the organometallic precursor from the interaction with carbonyl groups in PMMA [27,35], as also highlighted with FTIR results.

The analysis of cycle-dependent net mass gain can be used to highlight differences in mass uptake under constant conditions. In the plot reported in Figure 7b for different

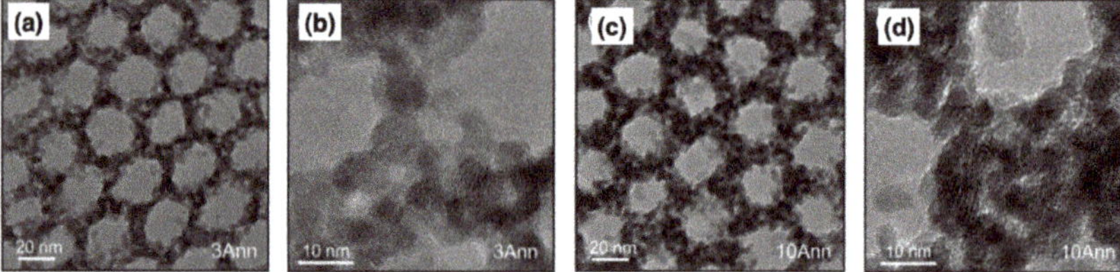

Figure 8. (**a–d**) TEM images at two different magnifications of the inorganic BCPs template, constituted of PS cylinders in a PMMA matrix infiltrated with (**a,b**) 3 cycles or (**c,d**) 10 cycles of In_2O_3. (**a–d**) are reproduced with permission from reference [64]. Copyright 2019 American Chemical Society.

STEM is a variation of conventional TEM in which a focused electron beam is raster scanned across the sample, previously thinned to allow transmission. Several detection modes are available giving STEM great versatility. On-axis detection of transmitted electrons yields bright-field intensity imaging, while the detection of fore scattered electrons complements it with annular dark-field (ADF) imaging, or high-angle annular dark-field (HAADF) imaging, giving Z-contrast information. Reference [51] reports the realization of heterostructure nanorod array through the simultaneous and spatially-controlled growth of Al_2O_3 and ZnO with a single SIS process in a BCPs film of PMMA cylinders in a PS matrix. HAADF STEM micrographs of the heterostructures acquired at different tilting angles are presented by the authors, showing contrast variation along the nanorods' length. EDX maps revealed the distribution of the target elements, Al and Zn, mainly at the top and bottom part of a nanostructure, respectively. In the same manuscript the authors also report a cross-sectional 3D reconstruction of the heterostructures, obtained by EDX-STEM tomography. Recently, HAAFD-STEM imaging was used to resolve the infiltrated ZnO at the junction of vertical and horizontal PLA in a three-dimensional structure of poly(1,1-dimethyl silacyclobutane-*b*-styrene-*b*-lactide) (PDMSB-*b*-PS-*b*-PLA) triblock terpolymer with PS and PLA blocks domains [61].

Atomic force microscopy (AFM) and, more generally, scanning probe microscopy (SPM) are microscopic methods for the topographic characterization of films and nanopatterned materials. The use of a scanning probe allows mapping the surface of the specimen with lateral and vertical resolution in the nanometer range. The characterization of polymers treated with SIS has been dedicated to monitoring the morphological evolution before and after the infiltration at different cycles, mostly on resist films treated for increased etch resistance in lithographic processes [65,66]. These measurements usually highlight an increase of the lateral size of the nanostructure, with consecutive reduction of their pitch, up to their complete merging, and rounded edges with increased number of cycles. Morphological analysis on self-assembled PMMA cylindrical nanodomains revealed swelling of the polymer and 25% increase in their lateral size after 5 SIS cycles, as reported in reference [67], consistently with SE observation in reference [56]. Additionally, compositional information may be retrieved from phase signal in tapping mode AFM measurements and nanomechanical properties may be investigated through force-distance measurements. Reference [67] reports PeakForce tapping mode for quantitative nanomechanical mapping (QNM) on SIS-treated homopolymers and self-assembled block copolymers. Young's modulus was monitored on the PMMA homopolymer layer and cylindrical nanodomains revealing an increased value after 5 and 11 SIS cycles, respectively, consistent with the incorporation of Al_2O_3 inside the polymer and increased stiffness. The results are reported in Figure 9a. Force-distance measurements on PMMA exhibited a decrease in the adhesion after infiltration with respect to the pristine polymer, as shown in Figure 9b. The same measurements on PS revealed no change in the stiffness or adhesion forces of the polymer.

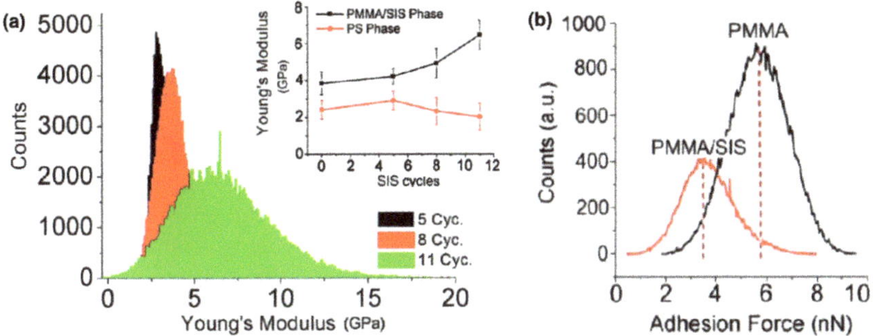

Figure 9. (**a**) Increase of the Young's modulus variation at 5, 8 and 11 SIS cycles in PMMA domains. In the inset, the variation of the Young's modulus for PMMA and PS phases is shown as a function of the number of SIS cycles. (**b**) Distribution of the adhesion force measured on PMMA nanodomains in before and after the infiltration process. All panels are reproduced with permission from reference [67]. Copyright 2017 American Chemical Society.

Time-of-flight secondary ions mass spectrometry (ToF-SIMS) is a destructive technique consisting in sputtering the material under study with a focused beam of primary ions. This generates secondary ions that pass through a time-of-flight mass spectrometer. When investigating polymeric samples, the use of bombarding ions clusters improves secondary ions yield and reduces damaging and molecular fragmentation [30]. The resulting composition, corresponding to different depths of the sputtering process and planar position of the rastering primary beam, is used to reconstruct the 3D cross section of the specimen, complementing the results from STEM and EDX spectroscopy. However, appropriate calibration standards are required for quantitative depth-profiling [68]. ToF-SIMS has been used to understand the depth distribution of oxides after SIS treatment, usually adopted in homopolymer layers such as PMMA [56,63] and PS [56], PET film and fibers [54], but also in block copolymers layers such as PS-*b*-P2VP both as micellar films [58] and self-assembled nanodomains infiltrated with SnO_x [43].

Thermogravimetric analysis (TGA), similarly to QCM gravimetry, yields information on the mass of infiltrated oxide in an *ex situ* process consisting in heating up the hybrid material and monitoring the weight change due to the loss of the polymeric volatile component. In reference [69], this technique confirmed the incorporation of alumina in polyethersulfone (PES) membrane with intact nanostructuration enabling the growth of laser-induced graphene (LIG). Among the techniques already mentioned for *in situ* phenomenological studies, FTIR and SE are also used in *ex situ* characterization. Attenuated total reflectance (ATR) FTIR, a variation of FTIR in reflection mode, has been reported in several works [62,69], including grazing incidence configuration [65], as a useful analytical method for cycle-dependent chemical characterization of the infiltrated polymer properties. Spectroscopic ellipsometry is often used in *ex situ* studies to measure the thickness variation of the polymer during the main processing steps (i.e., prior to SIS, after SIS, and after the polymer ashing) [62]. It can also provide information on the modified refractive index of the hybrid materials thus supporting application in optics and optoelectronics and related fields.

The characterization of the hybrid materials' properties after the SIS process is supported by several methods described so far, dedicated to chemical, morphological, mechanical, structural and optical analysis. Some of the most common techniques, such as STEM and EDX and ToF-SIMS analysis, present time-consuming preparation or destructive operations, compromising the functionality of the investigated materials. Another category of analytical methods, not yet mentioned in this review, is constituted by X-ray techniques allowing non-destructive versatile multidimensional investigation at high-resolution in both laboratory settings and synchrotron facilities. Structural properties can be characterized through X-ray diffraction (XRD), where the peaks' intensity and position in the diffraction pattern identify the atomic arrangement univocally yielding information on phase, crystallographic orientation, crystallinity, grain size, strain and defects. Local chemical and electronic structure around selected atomic species in a material can be retrieved through element-specific measurements of the first and second shell coordination distances by X-ray absorption spectroscopy (XAS) inner-shell methods. These rely on brilliant X-ray beams to probe the material with energies near the target element's adsorption edge or far above it for near-edge X-ray absorption fine structure (NEXAFS), also known as XANES, or extended X-ray absorption fine structure (EXAFS), respectively. Finally, morphological properties at the nanoscale can be studied through X-ray scattering (XRS) methods that display scattered photon intensities as a function of the momentum transfer Q $(1/\text{Å})$. Particularly, GISAXS, operated in grazing-incidence mode and analyzing small-angle scattering, is not new to the BCPs community and has been largely applied to study the nanoscale morphology of BCPs templates [70,71].

A notable multidimensional *ex situ* characterization using the former methods has been recently presented in reference [64] to study the atomic-scale structure and the possible mechanism of nucleation of TMIn precursor in PS-*b*-PMMA BCPs. Powder diffraction (PXRD) analysis of the crystalline structure of as-grown hybrid InO_xH_y/PMMA thin film, already shown in Figure 8a–d. The resulting XRD peaks exhibit high broadness indicating randomly distributed inorganic phase domains without long-range crystallographic order, compatible with an amorphous structure formed at low processing temperature (80 °C). Concurrently, EXAFS analysis was carried out on as-grown and annealed infiltrated PS-b-PMMA films showing a transition from InO_xH_y clusters to crystalline structures, whose local coordination environment after annealing was compatible with cubic In_2O_3 and $In(OH)_3$. In addition, high-energy X-ray scattering (HEXS) measurements have been paired with atomic pair distribution function analysis (PDF) and, in combination with EXAFS on annealed samples, confirmed the formation of an inorganic mesoporous film with sub-6 nm In_2O_3 cubic nanocrystals. HEXS-PDF analysis allows to retrieve the size of the inorganic clusters at each SIS cycle as well as their possible atomic structures [64,72].

Another noteworthy X-ray analytical method is X-ray photoelectron spectroscopy (XPS), also known as electron spectroscopy for chemical analysis (ESCA), is a common technique for surface chemistry analysis, usually implemented with laboratory setup. An X-ray beam impinges on the sample surface and generates photoelectrons at different energies. The energy spectrum enables the identification of the surface composition, chemical and electronic state. The characterization of infiltrated polymers is usually carried out *ex situ* to determine the chemical state of the oxide growth or chemical modification of the infiltrated polymer [33,43,58,65,69,73]. In reference [43], the authors report XPS measurements on SnO_x infiltrated in P2VP homopolymer films as shown in Figure 10a,b. XPS enables the identification of Sn $3d_{5/2}$ and Sn $3d_{3/2}$ transitions visible in the spectrum, indicating that both tin oxides with Sn(IV) and Sn(II) oxidation states can be grown in the polymer layer. In reference [69], XPS was adopted as evidence of the incorporation of alumina inside PES membranes through the identification of Al 2p intense peak after the SIS process. Other works presenting X-ray-based characterization of the BCPs properties include reference [74] where GISAXS has been implemented to characterize the time-dependent morphological evolution of the BCPs matrix during the SIS process and reference [73] combining XPS with GISAXS and X-ray reflectivity (XRR) to study surface active polymer additives in BCP formulations.

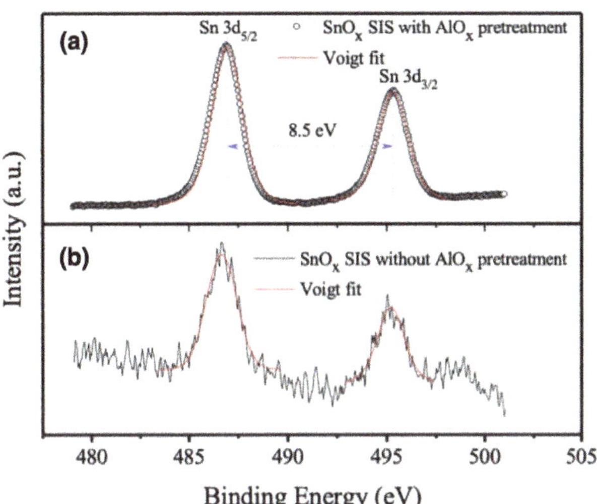

Figure 10. X-ray photoelectron spectroscopy (XPS) spectra recorded for Sn of SnO$_x$ grown by SIS with (a) pre-treatment and (b) without pre-treatment processing showing both Sn 3d$_{5/2}$ and Sn 3d$_{3/2}$ (P transitions. Adapted with permission from reference [43]. Copyright 2019 Elsevier Inc.

With respect to other methods, such as FTIR, QCM or TEM analysis, characterization through the previous and other X-ray methods has not yet reached a widespread diffusion in the SIS community despite these could provide a better understanding of the process–structure correlation. The encouraging straightforward and non-destructive acquisition is still associated with some challenges with regards to separating the organic and inorganic contributions of SIS complex and hybrid structures in X-ray scattering, reflectivity and spectroscopic signals [27,64,72].

4. Control of the Materials' Functional Properties by SIS

4.1. Optical Properties

The capability to selectively include metal oxide species inside self-assembled polymeric materials opened several opportunities in technological fields requiring the manipulation of light. A clear example is the realization of anti-reflective coatings (ARC) covering flat-panel displays of electronic devices, solar cells, curved optical elements or light-emitting diodes. To this goal, materials with refractive indices below 1.2 are required. To date, the literature describes two distinct approaches useful for the realization of BCPs-based ARC. The first approach relies on the increase of the absorption coefficient of incident light, occurring as a consequence of multiple reflections and scattering inside free-standing silicon nanopillars (SiNPs). In this context, the inclusion of metal oxides in ultrahigh molecular weight BCPs and the use of conventional reactive ion etching (RIE) processes allowed the formation of SiNPs with omnidirectional broadband anti-reflective capability (R < 0.16% in a wavelength range between 400 and 900 nm at an angle of incidence of 30°) [75]. A similar approach has been developed to obtain freestanding n-ZnO/p-Si nanotubes with low reflectivity in the UV-to-green light wavelength range (Figure 11a) [76]. The main drawback related to the use of SiNPs or nanotubes is the reduction of the transparency of the ARC film, strongly limiting the range of applications.

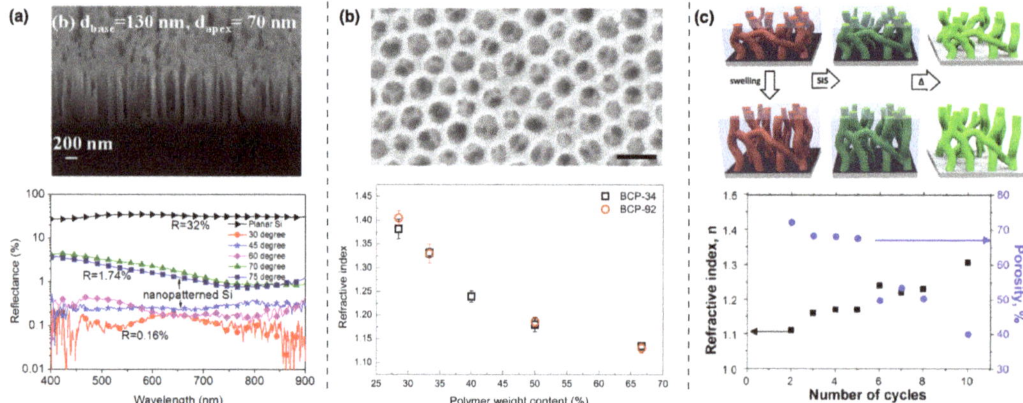

Figure 11. Broadband BCPs-based anti-reflective coatings (ARC) realized by (**a**) silicon nanopillars (**b**) TiO$_2$ nanocrystals inclusion inside poly(1,4-isoprene)-*block*-poly(ethylene oxide) (PI-*b*-PEO) micelles and (**c**) sequential infiltration synthesis of Al$_2$O$_3$ in cylindrical phase PS-*b*-P4VP. With these techniques, refractive index values approaching to $n_{ARC} \approx 1.1$ can be achieved. (**a**) Adapted with permission from reference [75]. Copyright 2017 American Chemical Society. (**b**) Adapted with permission from reference [77]. Copyright 2013 American Chemical Society. (**c**) Adapted with permission from reference [78]. Copyright 2017 American Chemical Society.

To extend the use of BCPs-based ARC to transparent substrates, the anti-reflective capabilities of an ARC can be tuned by adjusting its refractive index (n_{ARC}) and thickness (h_{ARC}), in such a way to induce destructive interference in the light reflected by the air/ARC and ARC/substrate interfaces. According to the Fresnel equations, for a given wavelength λ_0 and at a given angle of incidence, the best ARC conditions are accomplished for $n_{ARC} = \sqrt{n_{sub} \cdot n_{air}}$ (being n_{sub} and n_{air} the refractive of the substrate and air respectively) and $h_{ARC} \approx \lambda_0/4$, in the so-called "quarter-wave coatings". Following this approach, Guldin and coworkers [77] realized one of the first examples of BCPs-based ARC, exploiting a combination of silica-based sol-gel chemistry and preformed TiO$_2$ nanocrystals, selectively embedded inside poly(1,4-isoprene)-*block*-poly(ethylene oxide) (PI-*b*-PEO) micelles (Figure 11b). This type of composite materials combine the possibility of obtaining very low refractive index values (i.e., $n_{ARC} = 1.13$ at $\lambda_0 = 632$ nm) with self-cleaning properties. In fact, TiO$_2$-based photocatalysis can be used to degrade the hydrocarbons adsorbed on the ARC and restore its pristine anti-reflective properties.

In 2017 Berman et al. [78] proposed a novel method, the solvent-assisted SIS, as an efficient approach to create conformal coatings with very low n_{ARC} (Figure 11c) over a broad spectral range. With this method, the refractive index of inorganic coatings can be finely tailored by tuning the geometric parameters of the BCPs template (i.e., film thickens, swelling ratio, porosity, feature size and periodicity) as well as the deposition parameters (i.e., type of infiltrated material, number of cycles). As a result, the authors demonstrated that the refractive index of Al$_2$O$_3$ was lowered from 1.76 down to 1.10.

Beside the optical behavior linked to the change in refractive index, the nanostructured materials obtained by BCPs self-assembly and SIS exhibit interesting photoemission properties. Particular attention was paid to the electro- and photo-luminescence of nanostructures based on ZnO, a biocompatible and non-toxic material [79] with a wide range of potential applications in photonics [48,80], solid-state devices [81], gas sensors [82], water treatment [83] and biosensors [84].

The SIS process of zinc oxide is particularly complex, since the direct infiltration of diethylzinc (DEZ) precursor inside the polymer matrix often results in the formation of sparse ZnO nanoparticles [44,85]. For this reason a seeding treatment with a more reactive metal oxide (e.g., Al$_2$O$_3$) is often required. Ocola et al. demonstrated that the seeding treatment and the polymeric matrix strongly influence the emissive properties of the ZnO nanostructures [41]. Figure 12a,d show the variation of PL spectrum at the earlier stages of the SIS (i.e., Al$_2$O$_3$ seeding, first half cycle of DEZ, second half cycle of H$_2$O and second half cycle of DEZ). Dimer Zn atoms (O–Zn–O–Zn and O–Zn–O–Zn–O) provide strong UV and VIS photoluminescence emission, 20 times greater than that obtained from the mono Zn atoms (O–Zn and O–Zn–O). For an increasing number of SIS cycles the authors observed the formation of crystals and consequent suppression of the VIS component of the PL emission (Figure 12e). It is worth noting that in the infiltration process of ZnO inside the PMMA matrix, the polymer does not constitute a passive host matrix for the DEZ precursor, but actively contributes to the PL of the nanostructures. Evidence of energy transfer between the PMMA and ZnO were demonstrated, while micro- and nano-patterning of the PMMA allows the manipulation of the PL spectrum of the infiltrated ZnO [86]. The large variation of the luminescence spectrum of ZnO, as a function of the deposition parameters, type and shape of the host matrix, represents a strong limiting factor to its diffusion in photonic applications. In this context, the infiltration of ZnO inside self-assembled BCPs matrices represent a viable way to obtain well-defined and periodic arrays of nanoparticles or nanowires (NWs) with improved photoemission capabilities in terms of spectral shape and intensity. In particular localized defects in ZnO nanoparticles, randomly disposed by drop casting on a pre-patterned substrate, have been reported to be efficient electrically driven single photon sources, working at room temperature [87]. The deterministic positioning and reduction of the dispersion in size of ZnO nanoparticles, achieved by combining SIS and BCPs, allows for the integration in electro-optic devices, such as electrically driven optical resonators.

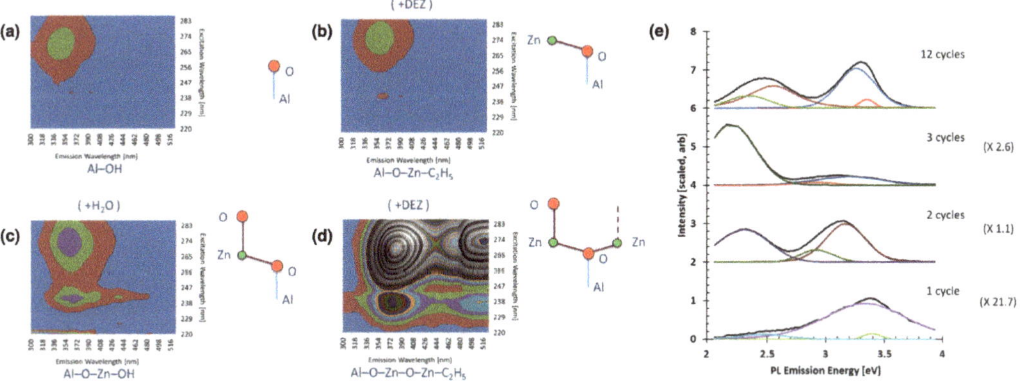

Figure 12. PL spectra recorded at different SIS steps and for variable excitation wavelengths between 220 and 285 nm: (**a**) Water terminated Al$_2$O$_3$ seed layer, (**b**) first half cycle of DEZ, (**c**) second half cycle of H$_2$O and (**d**) second half cycle of DEZ. The schematics on the right of all PL spectra illustrate the stage of ZnO growth that corresponds to each half cycle. (**e**) Emission spectra components as a function of the number of SIS cycles (the scaling factors are shown on the right side). All panels are adapted with permission from reference [41]. Copyright 2017 American Chemical Society.

4.2. Electrical Properties

The ability to control the level of doping of inorganic semiconductor materials has always driven the development of electronics. The same concept holds for the development of organic electronics where tailoring the doping level of organic functional materials is a prerequisite to control their electrical properties. With the growing interest in organic materials for printed and flexible electronics, light-emitting diodes (OLEDs), thin-film transistors, photovoltaic cells and batteries [88–93], several techniques based on the insertion of inorganic materials into polymers has been developed for the fabrication hybrid organic–inorganic materials with tailored electrical properties. Many of these processes alter polymer conductivity by doping with inorganic protonic acids, organic acids, Lewis acids, alkali metal salts or transition metal salts. These processes usually rely on wet chemistry with inherent limitations related to the solubility, temperature and can affect the polymer morphology, structure and purity [94–96]. In this scenario, SIS represents a solvent-free viable alternative to control electrical characteristics of polymers since infiltrated organometallic precursors lead to chemical reactions in the polymer to form hybrid materials with modified electrical properties. In 2015, Yu et al. [97] demonstrated that SIS represents a versatile doping strategy for engineering electrical properties of several functional polymers including polydimethylsiloxane (PDMS), polyimide (Kapton) and PMMA. The electrical properties of these polymers were tuned by infiltrating AlO_x molecules by SIS with TMA as a precursor. In the case of PDMS and Kapton, that always presents a negatively charged surface when contacted to other materials, it was observed that the AlO_x doping can significantly reduce the electron affinity of polymers due to the strong tendency of AlO_x molecules of repulsing electrons. Instead, if the host polymer possesses a strong tendency to repulse electrons as the AlO_x doping, as the case of PMMA, the effect of AlO_x doping is to enhance the positive charge density. By exploiting the different electron affinities of undoped and doped polymers, authors demonstrated the realization of triboelectric nanogenerators (TENGs) to convert mechanical energy into electricity. It is important to remark that, in this case, SIS was exploited as a technique for tuning bulk electrical properties since the diffusion of TMA was observed to be of $\approx 3\,\mu m$.

Among organic semiconductors, polyaniline (PANI) with its highly conjugated π delocalized molecular backbone is one of the most prominent conductive polymers finding applications in energy storage/conversion, supercapacitors, rechargeable batteries, fuel cells and water hydrolysis [98]. For all these applications, controlling the conductivity of PANI plays a crucial role. Besides depending on different oxidation states of the polymer (fully reduced leucoemeraldine, half oxidized emeraldine base and fully oxidized pernigraniline states) [98], the PANI conductivity can be modified through SIS doping. In 2017, Wang et al. [99] reported doping of PANI with metal chlorides by considering $MoCl_5$ and $SnCl_4$ precursors. In particular, it was observed that the conductivity of the infiltrated polymer (measured by means of four-probe techniques to avoid the effect of contact resistances) is correlated to the number of infiltration cycles and can be enhanced by up to six orders of magnitude. In the case of PANI infiltrated with $MoCl_5$, the highest conductivity of $2.93 \times 10^{-4}\,S\,cm^{-1}$ was observed after 100 infiltration cycles while in the case of PANI infiltrated with $SnCl_4$ the highest conductivity of $1.03 \times 10^{-5}\,S\,cm^{-1}$ was observed after 60 infiltration cycles (as a reference, untreated PANI shows conductivity $\leq 1 \times 10^{-10}\,S\,cm^{-1}$). Despite the conductivity of traditional HCl-doped PANI outperforms these results (doping with 1 M HCl results in conductivity of $8.23 \times 10^{-2}\,S\,cm^{-1}$), it was observed that metal chloride doped samples exhibited chemical stabilization, due to a much lower impact of the thermal treatments in vacuum on the doped polymer conductivity. In this case, the effect of doping was ascribed to the oxidation of the PANI and complexation of metal chlorides with the PANI nitrogen, with consequent enhancement of the electron mobility along the polymer chain.

A strong improvement of conductivity was reported also in the case of PANI infiltrated by ZnO using DEZ as a precursor, where mutual doping in between inorganic species and polymer constituents was achieved [96]. Indeed, in this case, the process was responsible for a reinforcement of the binding of ZnO to nitrogen of the polymer chain backbone inducing a Lewis-acid type of doping and, at the same time, for doping ZnO with nitrogen forming an interpenetrated network. As can be observed from Figure 13a, the number of infiltration cycles can be tuned to alter the PANI conductivity. In all cases, the conductivity is higher than the HCl-doped PANI. Also, since the exposure time is correlated with the infiltration depth, better conductivity performances were observed in the case of extended exposure times. Figure 13b reports conductivities of PANI doped with different infiltration parameters calculated from slopes of I–V characteristics. A maximum conductivity of $18.42\,S\,cm^{-1}$ was observed in the case of 600 infiltration cycles and 120 s of exposure time. It is worth noticing that the conductivity of the hybrid PANI/ZnO is a result of a synergy in between the involved materials since the conductivity is beyond the additive contribution of individual components. Indeed, lower conductivities were observed in the case of ALD-deposited ZnO films (refer to the conductivity represented by the green box of Figure 13b, where PANI was coated with an Al_2O_3 infiltration barrier before coating with an ALD-deposited ZnO). Similarly, W. Wang et al. [100] reported a VPI process to dope poly(3-hexyl)thiophene (P3HT) by means of the $MoCl_5$ precursor. In this case, the incorporation of Mo into the bulk polymer resulted in an increase of conductivity up to five order of magnitudes (a maximum of 3.01 S cm^{-1} was observed in the case of 100 infiltration cycles). In this case, changes in electrical conductivities are ascribed to a *p*-type doping related to the formation of a Lewis acid–base adduct formation between P3HT and $MoCl_5$, where P3HT acts as a Lewis base in conjunction with $MoCl_5$. In this framework, SIS results to be a promising strategy for solvent-free doping of polymers, making possible a top-down strategy to tune the electrical characteristics of pre-manufactured organic materials that can be implemented in roll-to-roll production lines for more efficient device fabrication of organic electronic devices. As a perspective, by properly selecting proper doping precursors and by controlling the infiltration conditions, the SIS strategy can be further explored for engineering electrical properties of a wide range of electrically conductive organic materials, where electrical characterization can be combined with UV-Vis, Raman, FTIR, XPS and XRD characterizations to understand chemical/structural changes of the polymer leading to a modification of its conduction properties.

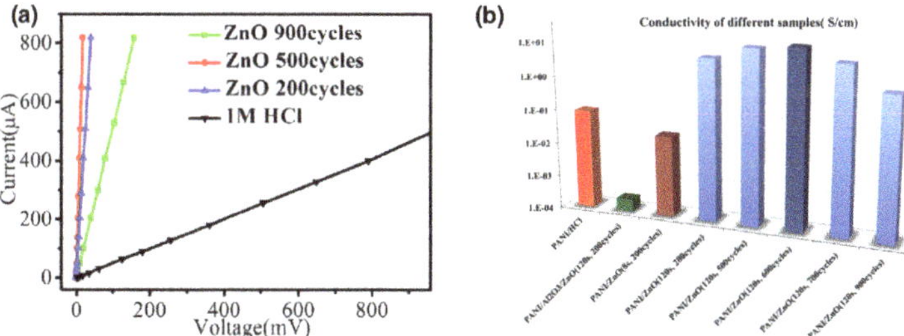

Figure 13. (**a**) I-V characteristics at room temperature of polyaniline (PANI) doped with different numbers of infiltration cycles (time exposure of 120 s). (**b**) Comparison of the conductivity of HCl-doped PANI (red box), atomic layer deposition (ALD)-deposited ZnO grown on PANI with an Al_2O_3 infiltration barrier (green box) and PANI infiltrated with ZnO by using different exposure time and cycle numbers. All panels are adapted with permission from reference [96]. Copyright 2017, American Chemical Society.

Infiltrated polymers can be exploited also for the realization of transparent and multifunctional sensors, as reported by Ocola et al. [101]. In particular, in their work it is reported that the SU-8 (usually employed as negative resist for lithographic purposes) infiltrated with ZnO can be exploited for the realization of highly sensitive UV sensors. However, a detailed understanding of the sensing mechanism relying on volume interactions of UV light with infiltrated polymers still needs further investigation. Also, SIS was demonstrated as a versatile technique for the realization of electrochemically stable conductive membranes. In their work, Bergsman et al. [69] reported that a SIS-based process enables the realization of conductive LIG coatings on porous polymer substrates. Indeed, the infiltration of PES membranes with alumina by using the TMA precursor is responsible for stabilization against deformation above the glass transition temperature of the polymer. This enables direct lasing of these polymeric membranes to form an LIG coating without affecting the membrane pore structure, allowing the realization of permeable conductive membranes (Figure 14a). Also, these membranes were observed to be electrochemically stable. The sheet resistance of SIS-treated LIG membranes evaluated by the Van der Pauw method was observed to be dependent on the laser power (Figure 14b) achieving the value of $(37.7 \pm 0.7)\ \Omega\ \square^{-1}$, a value that is comparable to the sheet resistance of carbon-nanotube

(CNT) composite materials. Note that without SIS treatment lased membranes exhibited an order of magnitude higher sheet resistance.

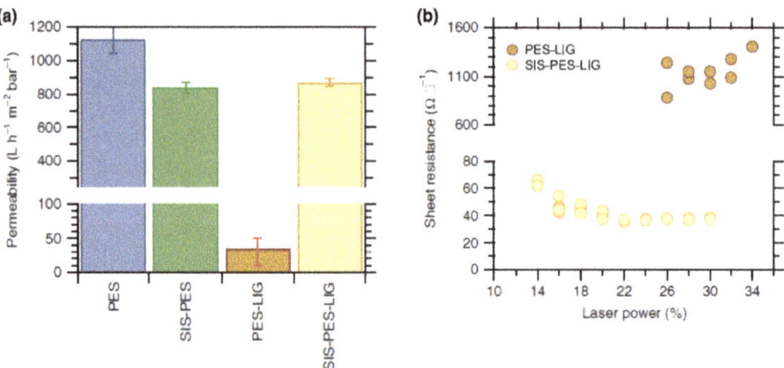

Figure 14. (**a**) Permeability of polyethersulfone (PES) membranes with and without SIS treatment before and after forming a laser-induced graphene (LIG) coating and (**b**) sheet resistance of lased membranes with and without SIS treatment as a function of the used laser power. All panels are reproduced and adapted under the terms of Creative Commons Attribution 4.0 License from reference [69]. Copyright 2020, the authors, published by Springer Nature.

The SIS technique was reported also as a versatile technique to grow semiconductive oxide thin films, as reported by Waldman et al. [49] that have synthesized In_2O_3 as a transparent conductive metal oxide. In their work, a process for growing In_2O_3 by using TMIn as a precursor and PMMA as substrate was established. After subsequent removal of PMMA and annealing at 400 °C, the remaining SIS-derived film exhibited typical electrical characteristics of undoped In_2O_3 thin films, as revealed by Hall effect measurements. Besides thin films, Vapor-phase infiltration can be exploited also for the realization of nanostructures based on metal oxides for the realization of electronic devices. For this purpose, the polymeric matrix can be patterned before the infiltration process in order to control position and geometries of nanostructures. In this framework, electrical properties of ZnO wires realized by means of SIS were investigated by Nam et al. [102]. As schematized in Figure 15a, the realization of ZnO stripes was performed by patterning a SU-8 template, subsequently infiltrated by ZnO and then removed by oxygen plasma. The resulting ZnO nanowires with length of 5 μm and width of about 50 nm present a nanocrystalline structure with grain sizes smaller than 5 nm. Subsequently, these nanostructures were contacted by means of source and drain contacts (Ti/Au) to realize an NW field effect transistor (NW-FET) device, exploiting the SiO_2 substrate as gate dielectric and Si as gate electrode (schematization in Figure 15b). Electrical characterization revealed that the ZnO NWs become semiconducting only after an annealing process at 500 °C for 10 min in hydrogen (4% H_2 with Ar balance) to increase carrier concentration. After that, the ZnO NW exhibited a *n*-type semiconducting behavior as can be observed from Figure 15c, where an increase of the gate voltage (V_G) resulted in an increase of the device conductivity. Similarly, an intrinsic *n*-type doping was reported in a wide range of ZnO nanostructures. It is worth noticing that a similar unintentional *n*-type doping was reported in a wide range of ZnO nanostructures and was ascribed to the presence of intrinsic defects and/or impurities that act as shallow donors [103]. Assuming the cylinder-on-plate model and by considering the transfer characteristics reported in Figure 15d, the carrier concentration was estimated to be at least 2.5×10^{19} cm^{-3} while the electron mobility was estimated to be about 0.07 cm^{-3}. It should be noticed that the here reported charge density results to be much larger than the charge density observed in the case of single-crystalline ZnO NWs grown with a bottom-up approach that was reported to be in the order of $10^{17} - 10^{18}$ cm^{-3} [104,105]. In order to achieve new insights into the electronic transport mechanism of ZnO NWs realized by means of SIS with a top-down approach and to compare results with single crystalline ZnO NWs realized with a bottom-up approach, temperature-dependent electrical characterizations are required.

Recently, it has also been demonstrated that SIS represents an inexpensive and scalable strategy for the realization of resistive switching memories (ReRAM) that is compatible with existing semiconductor nanofabrication methods and materials. Indeed, Chakrabatarti et al. [106] have shown that nanoporous AlO_x grown by infiltration of PMMA acts as a dielectric layer for ReRAM cells characterized by a high on/off ratio (>10^9), low switching voltages (about 600 mV), retention up to 10^4 s and pulsed endurance up to 1 million cycles. These characteristics make these cells promising for memory and neuromorphic applications.

Metal-oxide thin film nanoarchitectures can be also realized by combining SIS with self-assembled BCPs patterning exploited to generate nanomorphologies. By exploiting a MDIS protocol in hierarchical BCPs thin films, Subramanian et al. [31] reported the realization of three-dimensional ZnO nanomesh. Electrical conductivity across the multilayered nanomesh was observed to depend on the number of patterned layers. If a sufficient number of layers is realized, geometrical 3D charge percolation conduction is established across overlapping and orthogonal staking of nanowire fingerprint layers. For this reason, these systems represent percolative conduction networks where conductivity can be controlled by properly tuning geometrical parameters of the metal-oxide nanostructures. As a perspective, nanoarchitectures with tailored conductance properties can be realized by exploiting and combining different BCPs patterning strategies.

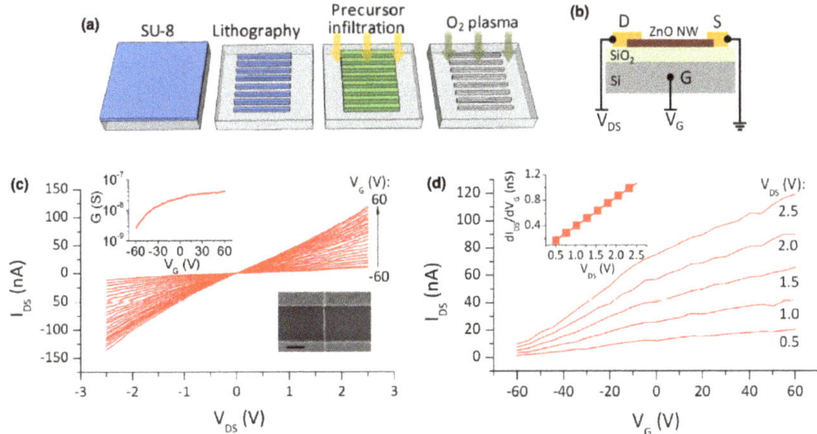

Figure 15. (**a**) Schematic representation of the ZnO patterning process consisting in the deposition of a SU-8 polymer, definition of SU-8 templates by lithography, infiltration synthesis with ZnO and formation of ZnO nanostructures by removing the polymer template through oxygen plasma. (**b**) Nanowire (NW) field effect transistor (NW-FET) transistor configuration where S, D and G represent source, drain and gate, respectively. (**c**) I_{DS} vs V_{DS} as a function of different V_G. The inset in the top left shows the dependence of the zero-bias conductance on V_G while the inset in the bottom right shows an SEM image of the NW-FET (scale bar of 500 nm). (**d**) I_{DS} vs V_G for different V_{DS}. The inset shows the dependence of the transconductance (d I_{DS} / d V_G) on V_{DS}. All panels are reproduced and adapted from reference [102]. Copyright 2015, AIP Publishing.

5. Conclusions and Perspectives

In recent years, a rapid expansion in SIS processing parameters has occurred [27]. Diverse vapor phase reactant combinations, pulses duration, purge duration, temperature and number of cycles have been tested on diverse polymers functional groups and block copolymers with varying Flory-Huggins parameter and molecular weight. The process kinetics and hybrid materials' properties have been probed through several analytical methods so far, constituting both a challenge and a push for progress. However, developing more and more reliable characterization methods is required to increase our knowledge and control capability on SIS when moving in the expanding process space. The basic metrological requirements must be met proceeding towards absolute quantitative methods and interlaboratory comparability. A great deal of information on the chemical and structural properties of SIS-processed BCPs is to be found in complementary approaches using *in situ* and *ex situ* optical, vibrational and X-ray spectroscopic methods in combination with more straightforward information from electron and scanning probe microscopy methods. The interpretation of characterization results may be supported through theoretical modeling and simulations, with density functional theory (DFT) being a prominent candidate to investigate the mechanism of chemical reactions and predict suitable conditions and reactants [52,107]. In this scenario, advancements in SIS are related to the development of a high throughput metrology at the nanoscale.

The correct interpretation of the chemical/physical mechanisms and precise characterization of the infiltrated BCPs are fundamental characteristics for the realization of photonic structures and electronic devices with improved functionalities. A clear example is the fabrication of nanostructured materials with non-linear optical properties (e.g., ZnO nanostructures) [108] or metamaterials (e.g., metal/dielectric hyperbolic metamaterials) [109]. Furthermore, advances in BCPs patterning and SIS techniques can be exploited for the realization of either electrodes and/or active materials of next-generation electronic devices to overcome obstacles of device downscaling and system integration. As an example, BCPs in conjunction with SIS can offer an efficient way for fabricating crossbar arrays of memristive devices for the realization of next-generation computing architectures for neuromorphic-type of data processing, in accordance with the roadmap on emerging hardware and technology for machine learning [20].

Artificial intelligence (AI) and machine learning techniques, already giving increasing contribution to the field of physical chemistry [110], can support experimental and theoretical work on SIS process parameters control and characterization [111] in order to design functional materials with tailorable properties to be exploited in optical, mechanical and electrical applications through a "materials by design" approach.

Author Contributions: Resources, E.C., I.M., G.M., F.F.L.; writing—original draft preparation, E.C., I.M., G.M., F.F.L.; writing—review and editing, E.C., I.M., G.M., N.D.L., L.B., F.F.L.; funding acquisition, N.D.L., L.B. All authors have read and agreed to the published version of the manuscript."

Funding: The project *16ENV07 Aeromet* has received funding from the EMPIR programme co-financed by the Participating States and from the European Union's Horizon 2020 research and innovation programme. The project *Volume Photography* received funding by the 2016 grant "Progetti premiali" of the Italian Ministry of University and Research.

Conflicts of Interest: The authors declare no conflict of interest.

References

1. Suh, H.S.; Moni, P.; Xiong, S.; Ocola, L.E.; Zaluzec, N.J.; Gleason, K.K.; Nealey, P.F. Sub-10-nm patterning via directed self-assembly of block copolymer films with a vapour-phase deposited topcoat. *Nat. Nanotechnol.* **2017**, *12*, 575–581.
2. Cummins, C.; Pino, G.; Mantione, D.; Fleury, G. Engineering block copolymer materials for patterning ultra-low dimensions. *Mol. Syst. Des. Eng.* **2020**, *5*, 1642–1657.
3. Ding, Y.; Gadelrab, K.R.; Rodriguez, K.M.; Huang, H.; Ross, C.A.; Alexander-Katz, A. Emergent symmetries in block copolymer epitaxy. *Nat. Commun.* **2019**, *10*, 1–7.
4. Stein, A.; Wright, G.; Yager, K.G.; Doerk, G.S.; Black, C.T. Selective directed self-assembly of coexisting morphologies using block copolymer blends. *Nat. Commun.* **2016**, *7*, 1–7.
5. Stefik, M.; Guldin, S.; Vignolini, S.; Wiesner, U.; Steiner, U. Block copolymer self-assembly for nanophotonics. *Chem. Soc. Rev.* **2015**, *44*, 5076–5091.
6. Yasen, W.; Dong, R.; Aini, A.; Zhu, X. Recent advances in supramolecular block copolymers for biomedical applications. *J. Mater. Chem. B* **2020**, *8*, 8219–8231.
7. Shiohara, A.; Prieto-Simon, B.; Voelcker, N.H. Porous polymeric membranes: Fabrication techniques and biomedical applications. *J. Mater. Chem. B* **2021**, *9*, 2129–2154.

8.	Orilall, M.C.; Wiesner, U. Block copolymer based composition and morphology control in nanostructured hybrid materials for energy conversion and storage: Solar cells, batteries, and fuel cells. *Chem. Soc. Rev.* **2011**, *40*, 520–535.

9.	Guo, C.; Lin, Y.H.; Witman, M.D.; Smith, K.A.; Wang, C.; Hexemer, A.; Strzalka, J.; Gomez, E.D.; Verduzco, R. Conjugated block copolymer photovoltaics with near 3% efficiency through microphase separation. *Nano Lett.* **2013**, *13*, 2957–2963.

10.	Liu, C.C.; Franke, E.; Mignot, Y.; Xie, R.; Yeung, C.W.; Zhang, J.; Chi, C.; Zhang, C.; Farrell, R.; Lai, K.; et al. Directed self-assembly of block copolymers for 7 nanometre FinFET technology and beyond. *Nat. Electron.* **2018**, *1*, 562–569.

11.	Jacobberger, R.M.; Thapar, V.; Wu, G.P.; Chang, T.H.; Saraswat, V.; Way, A.J.; Jinkins, K.R.; Ma, Z.; Nealey, P.F.; Hur, S.M.; et al. Boundary-directed epitaxy of block copolymers. *Nat. Commun.* **2020**, *11*, 1–10.

12.	Wu, M.L.; Wang, D.; Wan, L.J. Directed block copolymer self-assembly implemented via surface-embedded electrets. *Nat. Commun.* **2016**, *7*, 1–7.

13.	Ferrarese Lupi, F.; Giammaria, T.; Seguini, G.; Laus, M.; Enrico, E.; De Leo, N.; Boarino, L.; Ober, C.; Perego, M. Thermally induced orientational flipping of cylindrical phase diblock copolymers. *J. Mater. Chem. C* **2014**, *2*, 2175–2182.

14.	Giammaria, T.J.; Ferrarese Lupi, F.; Seguini, G.; Perego, M.; Vita, F.; Francescangeli, O.; Wenning, B.; Ober, C.K.; Sparnacci, K.; Antonioli, D.; et al. Micrometer-scale ordering of silicon-containing block copolymer thin films via high-temperature thermal treatments. *ACS Appl. Mater. Interfaces* **2016**, *8*, 9897–9908.

15.	Ferrarese Lupi, F.; Giammaria, T.J.; Miti, A.; Zuccheri, G.; Carignano, S.; Sparnacci, K.; Seguini, G.; De Leo, N.; Boarino, L.; Perego, M.; et al. Hierarchical order in dewetted block copolymer thin films on chemically patterned surfaces. *ACS Nano* **2018**, *12*, 7076–7085.

16.	Leniart, A.A.; Pula, P.; Sitkiewicz, A.; Majewski, P.W. Macroscopic Alignment of Block Copolymers on Silicon Substrates by Laser Annealing. *ACS Nano* **2020**, *14*, 4805–4815.

17.	Rahman, A.; Majewski, P.W.; Doerk, G.; Black, C.T.; Yager, K.G. Non-native three-dimensional block copolymer morphologies. *Nat. Commun.* **2016**, *7*, 1–8.

18.	Jiang, J.; Jacobs, A.G.; Wenning, B.; Liedel, C.; Thompson, M.O.; Ober, C.K. Ultrafast Self-Assembly of Sub-10 nm Block Copolymer Nanostructures by Solvent-Free High-Temperature Laser Annealing. *ACS Appl. Mater. Interfaces* **2017**, *9*, 31317–31324.

19.	Ferrarese Lupi, F.; Giammaria, T.J.; Seguini, G.; Vita, F.; Francescangeli, O.; Sparnacci, K.; Antonioli, D.; Gianotti, V.; Laus, M.; Perego, M. Fine tuning of lithographic masks through thin films of PS-b-PMMA with different molar mass by rapid thermal processing. *ACS Appl. Mater. Interfaces* **2014**, *6*, 7180–7188.

20.	Berggren, K.; Xia, Q.; Likharev, K.K.; Strukov, D.B.; Jiang, H.; Mikolajick, T.; Querlioz, D.; Salinga, M.; Erickson, J.R.; Pi, S.; et al. Roadmap on emerging hardware and technology for machine learning. *Nanotechnology* **2020**, *32*, 012002.

21.	Frascaroli, J.; Brivio, S.; Ferrarese Lupi, F.; Seguini, G.; Boarino, L.; Perego, M.; Spiga, S. Resistive switching in high-density nanodevices fabricated by block copolymer self-assembly. *ACS Nano* **2015**, *9*, 2518–2529.

22.	Murataj, I.; Channab, M.; Cara, E.; Pirri, C.F.; Boarino, L.; Angelini, A.; Ferrarese Lupi, F. Hyperbolic Metamaterials via Hierarchical Block Copolymer Nanostructures. *Adv. Opt. Mater.* **2020**, *2001933*, doi:10.21203/rs.3.rs-98929/v1.

23.	Kim, J.Y.; Kim, H.; Kim, B.H.; Chang, T.; Lim, J.; Jin, H.M.; Mun, J.H.; Choi, Y.J.; Chung, K.; Shin, J.; et al. Highly tunable refractive index visible-light metasurface from block copolymer self-assembly. *Nat. Commun.* **2016**, *7*, 1–9.

24.	She, Y.; Lee, J.; Diroll, B.T.; Lee, B.; Aouadi, S.; Shevchenko, E.V.; Berman, D. Rapid synthesis of nanoporous conformal coatings via plasma-enhanced sequential infiltration of a polymer template. *ACS Omega* **2017**, *2*, 7812–7819.

25.	Peng, Q.; Tseng, Y.C.; Darling, S.B.; Elam, J.W. A route to nanoscopic materials via sequential infiltration synthesis on block copolymer templates. *ACS Nano* **2011**, *5*, 4600–4606.

26.	Xu, J.; Berg, A.I.; Noheda, B.; Loos, K. Progress and perspective on polymer templating of multifunctional oxide nanostructures. *J. Appl. Phys.* **2020**, *128*, 190903.

27.	Waldman, R.Z.; Mandia, D.J.; Yanguas-Gil, A.; Martinson, A.B.; Elam, J.W.; Darling, S.B. The chemical physics of sequential infiltration synthesis—A thermodynamic and kinetic perspective. *J. Chem. Phys.* **2019**, *151*, 190901.

28.	Ingram, W.F.; Jur, J.S. Properties and applications of vapor infiltration into polymeric substrates. *JOM* **2019**, *71*, 238–245.

29.	McGuinness, E.K.; Zhang, F.; Ma, Y.; Lively, R.P.; Losego, M.D. Vapor phase infiltration of metal oxides into nanoporous polymers for organic solvent separation membranes. *Chem. Mater.* **2019**, *31*, 5509–5518.

30.	Leng, C.Z.; Losego, M.D. Vapor phase infiltration (VPI) for transforming polymers into organic–inorganic hybrid materials: A critical review of current progress and future challenges. *Mater. Horiz.* **2017**, *4*, 747–771.

31.	Subramanian, A.; Doerk, G.; Kisslinger, K.; Daniel, H.Y.; Grubbs, R.B.; Nam, C.Y. Three-dimensional electroactive ZnO nanomesh directly derived from hierarchically self-assembled block copolymer thin films. *Nanoscale* **2019**, *11*, 9533–9546.

32.	Lee, S.M.; Pippel, E.; Gösele, U.; Dresbach, C.; Qin, Y.; Chandran, C.V.; Bräuniger, T.; Hause, G.; Knez, M. Greatly increased toughness of infiltrated spider silk. *Science* **2009**, *324*, 488–492.

33.	Peng, Q.; Tseng, Y.C.; Long, Y.; Mane, A.U.; DiDona, S.; Darling, S.B.; Elam, J.W. Effect of nanostructured domains in self-assembled block copolymer films on sequential infiltration synthesis. *Langmuir* **2017**, *33*, 13214–13223.

34.	Dandley, E.C.; Needham, C.D.; Williams, P.S.; Brozena, A.H.; Oldham, C.J.; Parsons, G.N. Temperature-dependent reaction between trimethylaluminum and poly (methyl methacrylate) during sequential vapor infiltration: Experimental and ab initio analysis. *J. Mater. Chem. C* **2014**, *2*, 9416–9424.

35.	Biswas, M.; Libera, J.A.; Darling, S.B.; Elam, J.W. New insight into the mechanism of sequential infiltration synthesis from infrared spectroscopy. *Chem. Mater.* **2014**, *26*, 6135–6141.

36. Biswas, M.; Libera, J.A.; Darling, S.B.; Elam, J.W. Kinetics for the sequential infiltration synthesis of alumina in poly (methyl methacrylate): An infrared spectroscopic study. *J. Phys. Chem. C* **2015**, *119*, 14585–14592.

37. Caligiore, F.E.; Nazzari, D.; Cianci, E.; Sparnacci, K.; Laus, M.; Perego, M.; Seguini, G. Effect of the Density of Reactive Sites in P(S-r-MMA) Film during Al2O3 Growth by Sequential Infiltration Synthesis. *Adv. Mater. Interfaces* **2019**, *6*, 1900503.

38. Liapis, A.C.; Subramanian, A.; Cho, S.; Kisslinger, K.; Nam, C.Y.; Yun, S.H. Conformal Coating of Freestanding Particles by Vapor-Phase Infiltration. *Adv. Mater. Interfaces* **2020**, *7*, 2001323.

39. Hill, G.T.; Lee, D.T.; Williams, P.S.; Needham, C.D.; Dandley, E.C.; Oldham, C.J.; Parsons, G.N. Insight on the Sequential Vapor Infiltration Mechanisms of Trimethylaluminum with Poly (methyl methacrylate), Poly (vinylpyrrolidone), and Poly (acrylic acid). *J. Phys. Chem. C* **2019**, *123*, 16146–16152.

40. Biswas, M.; Libera, J.A.; Darling, S.B.; Elam, J.W. Polycaprolactone: A Promising Addition to the Sequential Infiltration Synthesis Polymer Family Identified through In Situ Infrared Spectroscopy. *ACS Appl. Polym. Mater.* **2020**, *2*, 5501–5510.

41. Ocola, L.E.; Connolly, A.; Gosztola, D.J.; Schaller, R.D.; Yanguas-Gil, A. Infiltrated zinc oxide in poly (methyl methacrylate): An atomic cycle growth study. *J. Phys. Chem. C* **2017**, *121*, 1893–1903.

42. Kim, J.J.; Suh, H.S.; Zhou, C.; Mane, A.U.; Lee, B.; Kim, S.; Emery, J.D.; Elam, J.W.; Nealey, P.F.; Fenter, P.; et al. Mechanistic understanding of tungsten oxide in-plane nanostructure growth via sequential infiltration synthesis. *Nanoscale* **2018**, *10*, 3469–3479.

43. Barick, B.K.; Simon, A.; Weisbord, I.; Shomrat, N.; Segal-Peretz, T. Tin oxide nanostructure fabrication via sequential infiltration synthesis in block copolymer thin films. *J. Colloid Interface Sci.* **2019**, *557*, 537–545.

44. Yi, D.H.; Nam, C.Y.; Doerk, G.; Black, C.T.; Grubbs, R.B. Infiltration synthesis of diverse metal oxide nanostructures from epoxidized diene–styrene block copolymer templates. *ACS Appl. Polym. Mater.* **2019**, *1*, 672–683.

45. Pilz, J.; Coclite, A.M.; Losego, M.D. Vapor phase infiltration of zinc oxide into thin films of cis-polyisoprene rubber. *Mater. Adv.* **2020**, *1*, 1695–1704.

46. She, Y.; Lee, J.; Lee, B.; Diroll, B.; Scharf, T.; Shevchenko, E.V.; Berman, D. Effect of the micelle opening in self-assembled amphiphilic block Co-polymer films on the infiltration of inorganic precursors. *Langmuir* **2019**, *35*, 796–803.

47. She, Y.; Goodman, E.D.; Lee, J.; Diroll, B.T.; Cargnello, M.; Shevchenko, E.V.; Berman, D. Block-co-polymer-assisted synthesis of all inorganic highly porous heterostructures with highly accessible thermally stable functional centers. *ACS Appl. Mater. Interfaces* **2019**, *11*, 30154–30162.

48. Peng, Q.; Tseng, Y.C.; Darling, S.B.; Elam, J.W. Nanoscopic patterned materials with tunable dimensions via atomic layer deposition on block copolymers. *Adv. Mater.* **2010**, *22*, 5129–5133.

49. Waldman, R.Z.; Jeon, N.; Mandia, D.J.; Heinonen, O.; Darling, S.B.; Martinson, A.B. Sequential infiltration synthesis of electronic materials: Group 13 oxides via metal alkyl precursors. *Chem. Mater.* **2019**, *31*, 5274–5285.

50. Berman, D.; Shevchenko, E. Design of functional composite and all-inorganic nanostructured materials via infiltration of polymer templates with inorganic precursors. *J. Mater. Chem. C* **2020**, *8*, 10604–10627.

51. Azoulay, R.; Shomrat, N.; Weisbord, I.; Atiya, G.; Segal-Peretz, T. Metal oxide heterostructure array via spatially controlled–growth within block copolymer templates. *Small* **2019**, *15*, 1904657.

52. Weisbord, I.; Shomrat, N.; Azoulay, R.; Kaushansky, A.; Segal-Peretz, T. Understanding and Controlling Polymer–Organometallic Precursor Interactions in Sequential Infiltration Synthesis. *Chem. Mater.* **2020**, *32*, 4499–4508.

53. Leng, C.Z.; Losego, M.D. A physiochemical processing kinetics model for the vapor phase infiltration of polymers: Measuring the energetics of precursor-polymer sorption, diffusion, and reaction. *Phys. Chem. Chem. Phys.* **2018**, *20*, 21506–21514.

54. Padbury, R.P.; Jur, J.S. Systematic study of trimethyl aluminum infiltration in polyethylene terephthalate and its effect on the mechanical properties of polyethylene terephthalate fibers. *J. Vac. Sci. Technol. A* **2015**, *33*, 01A112.

55. Young, M.J.; Choudhury, D.; Letourneau, S.; Mane, A.; Yanguas-Gil, A.; Elam, J.W. Molecular Layer Etching of Metalcone Films Using Lithium Organic Salts and Trimethylaluminum. *Chem. Mater.* **2020**, *32*, 992–1001.

56. Cianci, E.; Nazzari, D.; Seguini, G.; Perego, M. Trimethylaluminum diffusion in PMMA thin films during sequential infiltration synthesis: In situ dynamic spectroscopic ellipsometric investigation. *Adv. Mater. Interfaces* **2018**, *5*, 1801016.

57. Aprile, G.; Ferrarese Lupi, F.; Fretto, M.; Enrico, E.; De Leo, N.; Boarino, L.; Volpe, F.G.; Seguini, G.; Sparnacci, K.; Gianotti, V.; et al. Toward Lateral Length Standards at the Nanoscale Based on Diblock Copolymers. *ACS Appl. Mater. Interfaces* **2017**, *9*, 15685–15697.

58. Ishchenko, O.M.; Krishnamoorthy, S.; Valle, N.; Guillot, J.; Turek, P.; Fechete, I.; Lenoble, D. Investigating sequential vapor infiltration synthesis on block-copolymer-templated titania nanoarrays. *J. Phys. Chem. C* **2016**, *120*, 7067–7076.

59. Ozaki, Y.; Ito, S.; Hiroshiba, N.; Nakamura, T.; Nakagawa, M. Elemental depth profiles and plasma etching rates of positive-tone electron beam resists after sequential infiltration synthesis of alumina. *Jpn. J. Appl. Phys.* **2018**, *57*, 06HG01.

60. Segal-Peretz, T.; Winterstein, J.; Doxastakis, M.; Ramirez-Hernandez, A.; Biswas, M.; Ren, J.; Suh, H.S.; Darling, S.B.; Liddle, J.A.; Elam, J.W.; et al. Characterizing the three-dimensional structure of block copolymers via sequential infiltration synthesis and scanning transmission electron tomography. *ACS Nano* **2015**, *9*, 5333–5347.

61. Lee, S.; Subramanian, A.; Tiwale, N.; Kisslinger, K.; Mumtaz, M.; Shi, L.Y.; Aissou, K.; Nam, C.Y.; Ross, C.A. Resolving Triblock Terpolymer Morphologies by Vapor-Phase Infiltration. *Chem. Mater.* **2020**, *32*, 5309–5316.

62. McGuinness, E.K.; Leng, C.Z.; Losego, M.D. Increased Chemical Stability of Vapor-Phase Infiltrated AlO x–Poly (methyl methacrylate) Hybrid Materials. *ACS Appl. Polym. Mater.* **2020**, *2*, 1335–1344.

63. Ito, S.; Ozaki, Y.; Nakamura, T.; Nakagawa, M. Depth profiles of aluminum component in sequential infiltration synthesis-treated electron beam resist films analyzed by time-of-flight secondary ion mass spectrometry. *Jpn. J. Appl. Phys.* **2020**, *59*, SIIC03.

64. He, X.; Waldman, R.Z.; Mandia, D.J.; Jeon, N.; Zaluzec, N.J.; Borkiewicz, O.J.; Ruett, U.; Darling, S.B.; Martinson, A.B.; Tiede, D.M. Resolving the Atomic Structure of Sequential Infiltration Synthesis Derived Inorganic Clusters. *ACS Nano* **2020**, *14*, 14846–14860.

65. Marneffe, J.F.d.; Chan, B.T.; Spieser, M.; Vereecke, G.; Naumov, S.; Vanhaeren, D.; Wolf, H.; Knoll, A.W. Conversion of a patterned organic resist into a high performance inorganic hard mask for high resolution pattern transfer. *ACS Nano* **2018**, *12*, 11152–11160.

66. Ozaki, Y.; Ito, S.; Nakamura, T.; Nakagawa, M. Sequential infiltration synthesis-and solvent annealing-induced morphological changes in positive-tone e-beam resist patterns evaluated by atomic force microscopy. *Jpn. J. Appl. Phys.* **2019**, *58*, SDDJ04.

67. Lorenzoni, M.; Evangelio, L.; Fernandez-Regulez, M.; Nicolet, C.; Navarro, C.; Perez-Murano, F. Sequential infiltration of self-assembled block copolymers: A study by atomic force microscopy. *J. Phys. Chem. C* **2017**, *121*, 3078–3086.

68. Massonnet, P.; Heeren, R.M. A concise tutorial review of TOF-SIMS based molecular and cellular imaging. *J. Anal. At. Spectrom.* **2019**, *34*, 2217–2228.

69. Bergsman, D.S.; Getachew, B.A.; Cooper, C.B.; Grossman, J.C. Preserving nanoscale features in polymers during laser induced graphene formation using sequential infiltration synthesis. *Nat. Commun.* **2020**, *11*, 1–8.

70. Ferrarese Lupi, F.; Giammaria, T.J.; Seguini, G.; Laus, M.; Dubček, P.; Pivac, B.; Bernstorff, S.; Perego, M. GISAXS analysis of the in-depth morphology of thick PS-b-PMMA films. *ACS Appl. Mater. Interfaces* **2017**, *9*, 11054–11063.

71. Dialameh, M.; Ferrarese Lupi, F.; Hönicke, P.; Kayser, Y.; Beckhoff, B.; Weimann, T.; Fleischmann, C.; Vandervorst, W.; Dubček, P.; Pivac, B.; et al. Development and Synchrotron-Based Characterization of Al and Cr Nanostructures as Potential Calibration Samples for 3D Analytical Techniques. *Phys. Status Solidi A* **2018**, *215*, 1700866.

72. Tiede, D.M.; Kwon, G.; He, X.; Mulfort, K.L.; Martinson, A.B. Characterizing electronic and atomic structures for amorphous and molecular metal oxide catalysts at functional interfaces by combining soft X-ray spectroscopy and high-energy X-ray scattering. *Nanoscale* **2020**, *12*, 13276–13296.

73. Vora, A.; Schmidt, K.; Alva, G.; Arellano, N.; Magbitang, T.; Chunder, A.; Thompson, L.E.; Lofano, E.; Pitera, J.W.; Cheng, J.Y.; et al. Orientation control of block copolymers using surface active, phase-preferential additives. *ACS Appl. Mater. Interfaces* **2016**, *8*, 29808–29817.

74. Elam, J.W.; Biswas, M.; Darling, S.; Yanguas-Gil, A.; Emery, J.D.; Martinson, A.B.; Nealey, P.F.; Segal-Peretz, T.; Peng, Q.; Winterstein, J.; et al. New insights into sequential infiltration synthesis. *ECS Trans.* **2015**, *69*, 147.

75. Mokarian-Tabari, P.; Senthamaraikannan, R.; Glynn, C.; Collins, T.W.; Cummins, C.; Nugent, D.; O'Dwyer, C.; Morris, M.A. Large block copolymer self-assembly for fabrication of subwavelength nanostructures for applications in optics. *Nano Lett.* **2017**, *17*, 2973–2978.

76. Wan, Z.; Lee, H.J.; Kim, H.G.; Jo, G.C.; Park, W.I.; Ryu, S.W.; Lee, H.B.R.; Kwon, S.H. Circular Double-Patterning Lithography Using a Block Copolymer Template and Atomic Layer Deposition. *Adv. Mater. Interfaces* **2018**, *5*, 1800054.

77. Guldin, S.; Kohn, P.; Stefik, M.; Song, J.; Divitini, G.; Ecarla, F.; Ducati, C.; Wiesner, U.; Steiner, U. Self-cleaning antireflective optical coatings. *Nano Lett.* **2013**, *13*, 5329–5335.

78. Berman, D.; Guha, S.; Lee, B.; Elam, J.W.; Darling, S.B.; Shevchenko, E.V. Sequential infiltration synthesis for the design of low refractive index surface coatings with controllable thickness. *ACS Nano* **2017**, *11*, 2521–2530.

79. Li, Z.; Yang, R.; Yu, M.; Bai, F.; Li, C.; Wang, Z.L. Cellular level biocompatibility and biosafety of ZnO nanowires. *J. Phys. Chem. C* **2008**, *112*, 20114–20117.

80. Huang, K.M.; Ho, C.L.; Chang, H.J.; Wu, M.C. Fabrication of inverted zinc oxide photonic crystal using sol–gel solution by spin coating method. *Nanoscale Res. Lett.* **2013**, *8*, 306.

81. Pan, J.; Chen, J.; Huang, Q.; Khan, Q.; Liu, X.; Tao, Z.; Zhang, Z.; Lei, W.; Nathan, A. Size tunable ZnO nanoparticles to enhance electron injection in solution processed QLEDs. *ACS Photonics* **2016**, *3*, 215–222.

82. Zhu, L.; Zeng, W. Room-temperature gas sensing of ZnO-based gas sensor: A review. *Sens. Actuators A Phys.* **2017**, *267*, 242–261.

83. Mustapha, S.; Ndamitso, M.; Abdulkareem, A.; Tijani, J.; Shuaib, D.; Ajala, A.; Mohammed, A. Application of TiO_2 and ZnO nanoparticles immobilized on clay in wastewater treatment: A review. *Appl. Water Sci.* **2020**, *10*, 1–36.

84. Fang, B.; Zhang, C.; Wang, G.; Wang, M.; Ji, Y. A glucose oxidase immobilization platform for glucose biosensor using ZnO hollow nanospheres. *Sens. Actuators B Chem.* **2011**, *155*, 304–310.

85. Kamcev, J.; Germack, D.S.; Nykypanchuk, D.; Grubbs, R.B.; Nam, C.Y.; Black, C.T. Chemically enhancing block copolymers for block-selective synthesis of self-assembled metal oxide nanostructures. *ACS Nano* **2013**, *7*, 339–346.

86. Ocola, L.E.; Gosztola, D.J.; Yanguas-Gil, A.; Suh, H.S.; Connolly, A. Photoluminescence of sequential infiltration synthesized ZnO nanostructures. In *Quantum Sensing and Nano Electronics and Photonics XIII*; International Society for Optics and Photonics: Bellingham, WA, USA, 2016; Volume 9755, p. 97552C.

87. Choi, S.; Berhane, A.M.; Gentle, A.; Ton-That, C.; Phillips, M.R.; Aharonovich, I. Electroluminescence from localized defects in zinc oxide: Toward electrically driven single photon sources at room temperature. *ACS Appl. Mater. Interfaces* **2015**, *7*, 5619–5623.

88. Berggren, M.; Nilsson, D.; Robinson, N.D. Organic materials for printed electronics. *Nat. Mater.* **2007**, *6*, 3–5.

89. Dimitrakopoulos, C.D.; Malenfant, P.R. Organic thin film transistors for large area electronics. *Adv. Mater.* **2002**, *14*, 99–117.

90. Gross, M.; Müller, D.C.; Nothofer, H.G.; Scherf, U.; Neher, D.; Bräuchle, C.; Meerholz, K. Improving the performance of doped π-conjugated polymers for use in organic light-emitting diodes. *Nature* **2000**, *405*, 661–665.

91. Hains, A.W.; Liang, Z.; Woodhouse, M.A.; Gregg, B.A. Molecular semiconductors in organic photovoltaic cells. *Chem. Rev.* **2010**, *110*, 6689–6735.
92. Forrest, S.R. The path to ubiquitous and low-cost organic electronic appliances on plastic. *Nature* **2004**, *428*, 911–918.
93. Muench, S.; Wild, A.; Friebe, C.; Häupler, B.; Janoschka, T.; Schubert, U.S. Polymer-based organic batteries. *Chem. Rev.* **2016**, *116*, 9438–9484.
94. Gregorczyk, K.; Knez, M. Hybrid nanomaterials through molecular and atomic layer deposition: Top down, bottom up, and in-between approaches to new materials. *Prog. Mater. Sci.* **2016**, *75*, 1–37.
95. Sanchez, C.; Julián, B.; Belleville, P.; Popall, M. Applications of hybrid organic–inorganic nanocomposites. *J. Mater. Chem.* **2005**, *15*, 3559–3592.
96. Wang, W.; Chen, C.; Tollan, C.; Yang, F.; Beltran, M.; Qin, Y.; Knez, M. Conductive Polymer–Inorganic Hybrid Materials through Synergistic Mutual Doping of the Constituents. *ACS Appl. Mater. Interfaces* **2017**, *9*, 27964–27971.
97. Yu, Y.; Li, Z.; Wang, Y.; Gong, S.; Wang, X. Sequential infiltration synthesis of doped polymer films with tunable electrical properties for efficient triboelectric nanogenerator development. *Adv. Mater.* **2015**, *27*, 4938–4944.
98. Wang, H.; Lin, J.; Shen, Z.X. Polyaniline (PANi) based electrode materials for energy storage and conversion. *J. Sci. Adv. Mater. Dev.* **2016**, *1*, 225–255.
99. Wang, W.; Yang, F.; Chen, C.; Zhang, L.; Qin, Y.; Knez, M. Tuning the conductivity of polyaniline through doping by means of single precursor vapor phase infiltration. *Adv. Mater. Interfaces* **2017**, *4*, 1600806.
100. Wang, W.; Chen, C.; Tollan, C.; Yang, F.; Qin, Y.; Knez, M. Efficient and controllable vapor to solid doping of the polythiophene P3HT by low temperature vapor phase infiltration. *J. Mater. Chem. C* **2017**, *5*, 2686–2694.
101. Ocola, L.E.; Wang, Y.; Divan, R.; Chen, J. Multifunctional UV and gas sensors based on vertically nanostructured zinc oxide: Volume versus surface effect. *Sensors* **2019**, *19*, 2061.
102. Nam, C.Y.; Stein, A.; Kisslinger, K.; Black, C.T. Electrical and structural properties of ZnO synthesized via infiltration of lithographically defined polymer templates. *Appl. Phys. Lett.* **2015**, *107*, 203106.
103. Janotti, A.; Van de Walle, C.G. Fundamentals of zinc oxide as a semiconductor. *Rep. Prog. Phys.* **2009**, *72*, 126501.
104. Chiu, S.P.; Lin, Y.H.; Lin, J.J. Electrical conduction mechanisms in natively doped ZnO nanowires. *Nanotechnology* **2008**, *20*, 015203.
105. Milano, G.; D'Ortenzi, L.; Bejtka, K.; Ciubini, B.; Porro, S.; Boarino, L.; Ricciardi, C. Metal-insulator transition in single crystalline ZnO nanowires. *Nanotechnology* **2021**, *32*, 185202.
106. Chakrabarti, B.; Chan, H.; Alam, K.; Koneru, A.; Gage, T.E.; Ocola, L.E.; Divan, R.; Rosenmann, D.; Khanna, A.; Grisafe, B.; et al. Nanoporous Dielectric Resistive Memories Using Sequential Infiltration Synthesis. *ACS Nano* **2021**, *15*, 4155–4164.
107. Yang, F.; Brede, J.; Ablat, H.; Abadia, M.; Zhang, L.; Rogero, C.; Elliott, S.D.; Knez, M. Reversible and irreversible reactions of trimethylaluminum with common organic functional groups as a model for molecular layer deposition and vapor phase infiltration. *Adv. Mater. Interfaces* **2017**, *4*, 1700237.
108. Waszkowska, K.; Krupka, O.; Kharchenko, O.; Figà, V.; Smokal, V.; Kutsevol, N.; Sahraoui, B. Influence of ZnO nanoparticles on nonlinear optical properties. *Appl. Nanosci.* **2020**, *10*, 4977–4982.
109. Huo, P.; Zhang, S.; Liang, Y.; Lu, Y.; Xu, T. Hyperbolic Metamaterials: Hyperbolic Metamaterials and Metasurfaces: Fundamentals and Applications. *Adv. Opt. Mater.* **2019**, *7*, 1970054.
110. Prezhdo, O.V. Advancing Physical Chemistry with Machine Learning. *J. Phys. Chem. Lett.* **2020**, *11*, 9656–9658.
111. Butler, K.T.; Davies, D.W.; Cartwright, H.; Isayev, O.; Walsh, A. Machine learning for molecular and materials science. *Nature* **2018**, *559*, 547–555.

 *nanomaterials*

Article

Characterisation of the PS-PMMA Interfaces in Microphase Separated Block Copolymer Thin Films by Analytical (S)TEM

Julius Bürger [1,2,3], Vinay S. Kunnathully [1,2,3], Daniel Kool [1,2,3], Jörg K. N. Lindner [1,2,3] and Katharina Brassat [1,2,3,*]

[1]   'Nanostructuring, Nanoanalysis and Photonic Materials' Group, Department of Physics, Paderborn University, 33098 Paderborn, Germany; buergerj@mail.upb.de (J.B.); vinay.kunnathully.sathees.kumar@uni-paderborn.de (V.S.K.); dkool@mail.upb.de (D.K.); lindner@physik.upb.de (J.K.N.L.)
[2]   Institute of Lightweight Design with Hybrid Materials (ILH), 33098 Paderborn, Germany
[3]   Center for Optoelectronics and Photonics Paderborn (CeOPP), 33098 Paderborn, Germany
*   Correspondence: katharina.brassat@uni-paderborn.de

Received: 18 December 2019; Accepted: 9 January 2020; Published: 13 January 2020

Abstract: Block copolymer (BCP) self-assembly is a promising tool for next generation lithography as microphase separated polymer domains in thin films can act as templates for surface nanopatterning with sub-20 nm features. The replicated patterns can, however, only be as precise as their templates. Thus, the investigation of the morphology of polymer domains is of great importance. Commonly used analytical techniques (neutron scattering, scanning force microscopy) either lack spatial information or nanoscale resolution. Using advanced analytical (scanning) transmission electron microscopy ((S)TEM), we provide real space information on polymer domain morphology and interfaces between polystyrene (PS) and polymethylmethacrylate (PMMA) in cylinder- and lamellae-forming BCPs at highest resolution. This allows us to correlate the internal structure of polymer domains with line edge roughnesses, interface widths and domain sizes. STEM is employed for high-resolution imaging, electron energy loss spectroscopy and energy filtered TEM (EFTEM) spectroscopic imaging for material identification and EFTEM thickness mapping for visualisation of material densities at defects. The volume fraction of non-phase separated polymer species can be analysed by EFTEM. These methods give new insights into the morphology of polymer domains the exact knowledge of which will allow to improve pattern quality for nanolithography.

Keywords: block copolymers; self-assembly; polymer interface; nanostructure metrology; line edge roughness LER; (S)TEM; STEM-EELS of PS and PMMA

---

1. Introduction

The self-assembly of block copolymers (BCPs) in thin films is one of the most promising approaches for next generation surface nanopatterning. A large variety of patterns with nanoscale features is accessible and can be easily created on large areas [1–5]. Block copolymer self-assembly is mainly driven by interfacial energies, i.e., polymer-polymer interactions of the BCP species and their interactions with a substrate or gaseous environment. During microphase separation, periodical arrays of sub-20 nm polymer domains are formed. The polymer domain shapes exhibit spherical, lamellar or cylindrical geometries determined by the polymer block length ratio [6–10]. These self-assembled polymer domains can then be used as templates for various purposes: removing one species selectively, one can create shadow masks for further lithographical processing [11], membranes used in nanofiltration [12,13] or versatile electrochemical devices [14]; Janus-type nanostructures can be created by microphase separation of terpolymers [15]; chemically functional polymers are exploited for domain specific diffusion for battery applications [16,17] and photovoltaics [18] or in sequential infiltration synthesis (SIS) processes [19,20]. For all these purposes the resulting pattern quality and device performance is determined by the initial morphology of the self-assembled polymer domains, i.e., the domain orientation within the film, the domain order and long range order as well as the accuracy of domain shapes.

The domain morphology is largely determined by the interface between the phase separated polymer domains. This interface between the polymer species is not sharp but interpenetration of polymer chains leads to concentration gradients. The resulting width of this interface is not negligible as it can easily exceed 40% of the actual domain size [21]. This interfacial width also strongly influences e.g., nanopattern line edge and line width roughness, which are important parameters for technological applications. The ratio between the equilibrium pattern periodicity L_0 and interfacial width Δ is discussed in recent literature as being crucial to estimate the suitability of a certain BCP system for sub-10 nm L_0 nanopatterning applications [21]. Thus, efforts are made to investigate the morphology of these interfaces and the origin of interfacial fluctuations in order to minimise interfacial widths [21–24]. New polymer species are, for instance, designed to reduce the interfacial width to <10% of L_0 [21].

The importance of the interface morphology for understanding polymer (de)mixing is being discussed since decades. The first model aiming to describe the interface between two polymer species was introduced by Helfand [25]. He investigated the interfacial tension $\square$ and interfacial width Δ by mean field theory for polymers within the strong segregation limit, showing that the interface is basically determined by the Flory Huggins parameter χ and the statistical polymer segment length a:

$$\square_\infty = k_B T \, a \, (\chi/6)^{1/2}/v \tag{1}$$

$$\Delta_\infty = 2a \, (6\chi)^{-1/2} \tag{2}$$

with v being the average monomeric volume. Similar to many other early predictions of the domain and interface morphology, this estimate assumes sufficiently large polymerisation degrees. By approximating infinite molecular weight (indicated by the subscripes ∞), effects of chain size and chain ends can be neglected [6,7,26,27]. It is to note that the interfaces between domains of block copolymers and respective blends of homopolymers are identical as was stated by the narrow-interface-approximation theory by Helfand and Wasserman [28] and experimentally verified by Shull et al. [29,30].

Later, finite-molecular weight and chain-end effects were found to decrease the interfacial tension and, thus, to increase the interfacial width [31,32]. An extended model including the chain length was proposed by Semenov in 1993 [33,34]. Since then, intensive further work on pitch scaling and interfacial width scaling in dependence of the effective Flory Huggins parameter χN (with N the degree of polymerisation) followed [35–40]. Simulations based on self-consistent field theory by C. T. Black [34] as well as A. Hannon and J. Kline [41] describe the interface width in good agreement with experimental observations

$$\Delta_x = \Delta_\infty \{1 + [24/(\chi N \pi^2)]^{1/3}\} \tag{3}$$

The broadening of the experimentally observed interface morphology compared to the simple thermodynamic model by Helfand is discussed to result from local fluctuations in the position of interfaces, thermal fluctuations of the concentration profiles, local stretching of the polymer chains [42,43] and polydispersity of polymer chains [44]. Semenov [33] describes these fluctuations as the deviation of interface positions σ in dependence of the interfacial tension, interfacial width and the polymer pattern periodicity L_0:

$$\sigma^2 = (2\pi\square_\infty)^{-1} \ln(L_0 \Delta_\infty^{-1}) \tag{4}$$

More recently, the interfacial width was found to be, in addition, dependent on the block copolymer annealing method [22] and the annealing temperature. The origin of these connections is the temperature dependency of the Flory Huggins parameter. It was observed that annealing near the order-disorder transition (ODT) temperature results in a larger interfacial width as the two blocks begin to mix, while the interface is sharper at annealing temperatures far below ODT. Therefore, the interfacial width is assumed to be a suitable measure for the progression of the phase separation [45].

The experimental observation and characterisation of the polymer domain morphology is demanding. Reciprocal space methods are being used to analyse order and polymer domain interfaces [45]. Measurements of X-ray and neutron reflectivity and evaluation of volume fraction profiles of the polymer species [22,40–42,46–50] or secondary ion mass spectrometry [51] are most common techniques. These methods, however, do not allow for acquisition of spatially resolved information and, for instance, neutron scattering experiments require deuteration of polymers to distinguish between the organic materials, which for many polymers is not easy in terms of synthesis and might influence the polymer behaviour. More recently resonant soft X-ray reflectivity (RSoXR) was introduced as a new method allowing for good contrast in unmodified organic materials [23,51–54]. For instance, recent work by Kline and coworkers [21] successfully applied RSoXR to determine domain periodicity and interfacial widths in high-χ block copolymers.

Real space analysis of block copolymer thin films is most commonly performed by scanning electron microscopy (SEM) and atomic force microscopy (AFM) [45]. These methods allow for the investigation of pattern order, however, resolution is limited and the interface between domains is not accessible. Transmission electron microscopy (TEM) of microphase separated BCPs is much more rarely used even though it is particularly suitable to investigate the morphology and shape of polymer domains at much higher resolution. Insightful works were published investigating polymer domain morphologies [55–57] and concentration profiles at polymer interfaces [44]. Recently, Segal-Peretz et al. [58,59] investigated morphologies and positional interface fluctuations of infiltrated polymer domains employing TEM tomography. Staining of one polymer domain e.g., with RuO_4 or OsO_4 [55–57] or infiltration of one polymer domain with e.g., Al_2O_3 [58] is most often used for contrast enhancement in these TEM studies. However, the incorporation of material for contrast improvement holds several disadvantages. It always leaves doubt on artefacts introduced with the foreign material due to chemical (cross-linking or chain scission reactions) or physical modifications (contraction or expansion of domains) [44]. It is also not fully understood how and where materials are infiltrated at domain interfaces where polymer chains can interpenetrate or form a concentration gradient [58]: Infiltration might reconstruct a material domain up to a certain threshold concentration or within all material volume containing any fraction of the distinct material. This will largely influence the apparent domain size and shape. In addition, selective staining is not easily available for all kinds of polymers.

To our knowledge, no high-resolution real space imaging of unmodified self-assembled polymer domains in block copolymer thin films has been published so far. This is probably due to many factors, including the difficulty to obtain reasonable contrast between the polymer species at electron energies above 100 keV where in the past TEMs used to have sufficient resolution, and the sensitivity of polymers to irradiation with energetic electrons. Attempts have been made to exploit phase contrasts induced between polymers by using a strong objective lens defocus, however, on the expense of spatial resolution [60]. Current analytical electron microscopes, equipped with correctors for spherical lens aberrations, and fast detectors can overcome these limits. Thus, in this work, we investigate the domain morphology and interface of unmodified unstained microphase separated polystyrene-*b*-polymethylmethacrylate (PS-*b*-PMMA) thin films by analytical (S)TEM. We investigate PS-*b*-PMMA BCPs with different block length ratios forming PMMA cylinders in PS (PS:PMMA 70:30), PS cylinders in PMMA (PS:PMMA 30:70) or alternating PS and PMMA lamellae (PS:PMMA 50:50).

We characterise the interface between PS and PMMA nanodomains in the block copolymers and correlate data from theory and literature on domain size, interface position and interfacial width to high resolution real space images. In particular, we image the polymer domain morphology and their interface using STEM revealing the internal structure of polymer domains, the positional fluctuation of the interfaces as well as the occurrence of grains enlarging the line edge roughness—all features which are not accessible by other techniques. STEM electron energy loss spectroscopy is used to investigate the chemical composition of polymer domains exploiting plasmonic resonances of the polymers as well as the near edge fine structure at their carbon K-edge. High-resolution imaging at the plasmon resonance and zero loss imaging are applied using energy filtered TEM spectroscopic imaging. Finally, energy filtered TEM thickness maps are shown to visualise the periodical density variations of the polymer domains revealing e.g., the polymer composition at defects in the polymer pattern.

2. Materials and Methods

2.1. Materials and Sample Preparation

Three different polystyrene-*b*-polymethylmethacrylate (PS-*b*-PMMA) block copolymers with block length ratios of PS:PMMA of 70:30, 30:70 and 50:50 were purchased from Polymer Source Inc. and dissolved in toluene (analytical grade, *C. Roth GmbH*). The molecular weights of the different polymers can be found in Table 1. The polydispersity indices are between 1.06–1.09. Thin films of the different block copolymers were spin casted onto silicon wafers covered with a thermally grown 700 nm thick silicon oxide sacrificial layer. For all BCPs, the oxide was functionalized with 5–7 nm thick PS-*co*-PMMA random copolymer brushes (M_n = 5.2–8.5 kg/mol, 58–66 mol% PS content) from Polymer Source Inc. BCP films with thicknesses of 35 nm for both, the PS:PMMA 70:30 and the PS:PMMA 30:70 BCP, and of 40 nm for the PS:PMMA 50:50 BCP were thermally annealed at 180 °C at a pressure of 10^{-7} mbar for 24 h to enable microphase separation. The long annealing time was chosen to ensure complete microphase separation. It was shown that the initial phase separation is an extremely rapid process which is only followed by a slow pattern optimisation through defect annihilation [45]. The long annealing of 24 h, thus, should allow for a terminated phase separation.

To obtain free-standing BCP membranes for TEM analysis, sample preparation was performed as described previously [11]. Briefly, microphase separated BCP films were released from the substrate by etching of the thick sacrificial silicon oxide layer in 10% HF_{aq} at room temperature. The floating BCP membrane was then skimmed off the etchant with a TEM grid. In this work, Quantifoil on Au grids purchased from Plano GmbH were used. Comparison of AFM images taken from the BCP film prior to and after HF dipping (not shown here) confirmed that the diluted HF solution of moderate temperature does not affect the polymers.

Table 1. Overview of polymer specifics. Molecular weights M_n (PS-PMMA), polymerisation degrees N_{PS}-N_{PMMA}, product of Flory Huggins parameter and polymerisation degree χN, domain size d_{SEM} and periodicity $L_{0,SEM}$ determined from SEM images, domain size d_{TEM} determined by TEM, height differences between polymer domains h_{AFM} determined by atomic force microscopy (AFM) (with polymethylmethacrylate (PMMA) domains exhibiting a larger thickness), interfacial widths Δ_x after Hannon/Kline and positional fluctuation of interface σ.

Polymer	M_n (PS-PMMA) [kg/mol]	N_{PS}-N_{PMMA}	χN	d_{SEM} [nm]	$L_{0,SEM}$ [nm]	d_{TEM} [nm]	h_{AFM} [nm]	Δ_x [nm]	σ [nm]
PS:PMMA 70:30	46.1–21.0	443–210	23.9	15.0± 2.6 [a]	35.0 ± 4.4	16.0 ± 0.9 [a]	1.12 ± 0.30	4.39	1.04
PS:PMMA 30:70	20.2–50.5	194–505	25.6	24.6± 4.5 [b]	35.6 ± 6.0	29.7 ± 0.6 [b]	1.15 ± 0.27	4.36	1.04
PS:PMMA 50:50	25.0–26.0	240–257	18.2	14.7 ± 2.4 [c]	24.3 ± 1.2	9.2 [c]	1.28 ± 0.28	4.53	0.93

[a] Diameter of PMMA cylinder; [b] diameter of PS cylinder; [c] width of PS lamella.

2.2. Characterisation Techniques

Scanning electron microscopy (SEM) images were taken with a Zeiss ultra plus at an acceleration voltage of 2 kV with an in-lens detector. Atomic force microscopy (AFM) was performed using a Digital Instruments Dimension 3100 in non-contact mode with 65 kHz Al-coated cantilevers (MikroMasch) with nominally 8 nm tip diameter.

Analytical (scanning) transmission electron microscopy ((S)TEM) was performed using a JEOL JEM-ARM200F equipped with a cold field emission electron gun (CFEG) and a probe-side mounted

ASCOR C_s-corrector (CEOS GmbH) allowing for the correction of geometric aberrations up to the 5th order and therefore resolution in the STEM mode of better than 1 Å at an acceleration voltage as low as 60 kV. All images shown in this work are acquired at this voltage where no beam damage, i.e., chemical or structural changes due to high-energy radiation, was observed. For comparison, images obtained at 200 kV can be found in the Supplementary Information Figure S1. All TEM images are acquired with 150 µm condenser lens aperture and captured on a 4 k × 4 k GATAN OneView camera. STEM images are recorded with a convergence semi-angle of 16.6 mrad on an annular dark field (ADF) detector, which collects scattered electrons at polar collection semi-angles of 69–147 mrad, using a 40 µm condenser lens aperture and a camera length of 12 cm.

Analytical methods such as energy filtered TEM (EFTEM) spectroscopic imaging (EFTEM-SI) and electron energy loss spectroscopy (EELS) are applied for differentiation between PS and PMMA domains. EFTEM and EELS are conducted with a GATAN GIF-Quantum ER image filter and are captured on a 2 k × 2 k CCD camera (GATAN UltraScan). Operating the CFEG at maximum beam current, the energy resolution as expressed by the zero-loss peak FWHM was 0.65–1.1 eV at 60 keV. Overview EEL spectra are generated with a broad beam in TEM mode, for spatially resolved analysis STEM-EELS line-scans were performed. EELS spectra are recorded in dual EELS mode with a dispersion of 0.1 eV for low-loss (−20 eV to 184.8 eV) and high-loss (200 eV to 404.8 eV) parts of the spectra. EFTEM thickness maps are obtained using the t/λ-method and mean free paths λ, calculated according to Iakoubovskii et al. [61].

2.3. Calculation of Expected Interfacial Widths and Interface Position Fluctuation

The theoretically expected interfacial widths for the polymers used in this work were determined using the models by Helfand and Kline, respectively. The Kuhn lengths, i.e., the statistical polymer segment lengths, of PS and PMMA are both approximately a = 0.70 nm [34].

As stated above, the Flory Huggins parameter is temperature dependent following

$$\chi(T) = \chi_S + \chi_H/T \tag{5}$$

with χ_S and χ_H being the entropic and enthalpic terms of the Flory Huggins parameter. For PS-PMMA these contributions were determined [62] to

$$\chi_{PS\text{-}PMMA} - 0.028 + 3.9/T \tag{6}$$

In this work, the Flory Huggins parameter for a temperature of 180 °C is used as this is the annealing temperature enabling microphase separation in the presented experiments.

The minimum interfacial width for a PS-PMMA interface according to the model by Helfand (Equation (2)) can be determined to

$$\Delta_\infty = 2.99 \text{ nm}$$

The interfacial widths according to the model by Kline et al. (Equation (3)) for the three PS-*b*-PMMA block copolymers used in this work are collected in Table 1. This model takes the polymerisation degree of the polymer blocks into account. Thus, the resulting interfacial width is larger than predicted by Helfand. However, as the polymerisation degrees and molecular weights of the three polymers are comparable, the expected interfacial widths only exhibit small differences. This is also in accordance to experimental investigations applying neutron scattering by Anastadiadis [46] who found similar interface widths at interfaces of polymers with molecular weights between M_n = 30–300 kg/mol.

The deviation of the interface position σ was determined according to Equation (4) with the interfacial tension $\square_\infty$ according to Equation (1) and the interfacial width Δ_x after the model by

Kline et al. (Equation (3). For simplification, the monomeric volume v was related to the statistical segment length a assuming that the segments occupy a spherical volume:

$$v = 0.52\ a^3$$

The positional fluctuation σ is usually directly correlated to the line edge roughness (LER). The results are assembled in Table 1.

3. Results

Figure 1 gives an overview of the three PS-*b*-PMMA block copolymers investigated in this work. Each row shows a different polymer with a block length ratio of PS:PMMA 70:30 (a–c), PS:PMMA 30:70 (d–f) or PS:PMMA 50:50 (g–i), respectively, investigated by scanning electron microscopy (SEM, left column), bright-field transmission electron microscopy (TEM, middle column) and atomic force microscopy (AFM) (right column). In all images both microphase separated PS and PMMA domains are shown after annealing, no polymer species was selectively removed or modified. SEM and AFM images were recorded with the BCP films supported by the substrate, while for TEM free standing membranes were used.

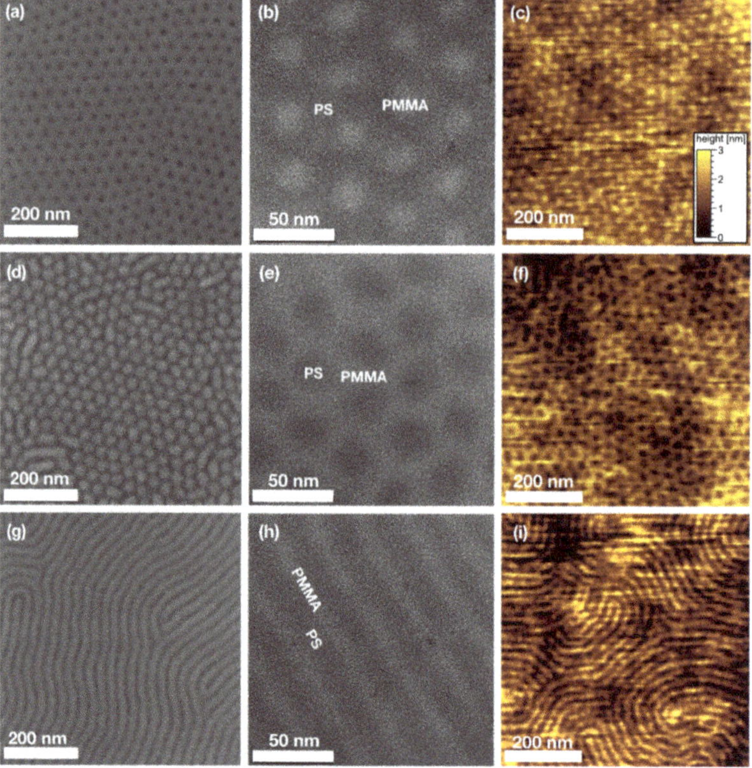

Figure 1. SEM, AFM and TEM images of the three polystyrene (PS)-b-PMMA block copolymers (BCPs) with block length ratios of (**a–c**) PS:PMMA 70:30, (**d–f**) PS:PMMA 30:70 and (**g–i**) PS:PMMA 50:50, respectively. (**a,d,g**) show SEM images, (**b,e,h**) display bright field TEM images and (**c,f,i**) are AFM height images with same height scales as in (**c**). In all images, both polymer species are apparent and no staining or other sample treatment was used.

SEM and AFM are the most commonly used techniques to analyse block copolymers. Domain size d and pattern periodicity L_0 are commonly determined from SEM images using grey scale thresholding techniques and image analysis based on Delaunay triangulation [5,63]. For the polymers investigated here, domain sizes and periodicities derived from SEM images are collected in Table 1. AFM imaging allows to determine height differences between the polymer domains. Figure 1c,f,i shows elevated PMMA domains with a difference of 1.1–1.3 nm to the PS domains.

TEM bright-field images of the three BCPs are shown for comparison. A contrast between PS and PMMA domains is visible in these unstained polymers, even though chemical compositions and densities of PS and PMMA are similar. While the resolution of SEM hardly allows to judge on the shape and lateral extension of polymer interfaces and AFM imaging is limited by the cantilever dimension, high-resolution TEM allows to investigate the polymer domain morphology close to the atomic scale. Thus, the exact shape of polymer domains as well as the blurred interface between distinct polymer domains become visible only here. This advantage will be employed in detail in the following using analytical TEM as well as STEM.

3.1. Polymer Domain Morphology and Line Edge Roughness Investigated by STEM-ADF

Figure 2 presents scanning transmission electron microscopy (STEM) images of the three unstained PS-*b*-PMMA block copolymers acquired with an annular dark field (ADF) detector. Dark areas correspond to PMMA domains while PS gives a bright contrast. Figure 2a shows PMMA cylinders in a PS matrix formed by BCP PS:PMMA 70:30, Figure 2b shows the inverse pattern, i.e., PS cylinders in a PMMA matrix (BCP PS:PMMA 30:70) and Figure 2c displays alternating PS and PMMA lamellae formed by the BCP PS:PMMA 50:50. It is to note that these contrasts, PS appearing brighter than PMMA, are not as expected. Heavier atoms should appear bright in dark-field imaging, thus, PMMA would be expected to give a brighter contrast compared to PS. The observed material assignment was, however, also reported by others [58] and is further verified by EFTEM thickness and elemental mapping (Sections 3.4 and 3.5).

The STEM-ADF images clearly show that the interface between PS and PMMA domains is not sharp but exhibits a broad, blurred material contrast between polymer species. The absence of sharp interfaces between polymer species in the STEM images is mainly due to the absence of sharp interfaces in the BCP film. The crucial influence of the imaging technique on the interpretation and analysis of the polymer domain morphology becomes apparent when comparing feature sizes determined from SEM and STEM-ADF images. In order to highlight this difference sketches of the domain sizes as measured from SEM images of these exact polymers are added in Figure 2a–c as white dashed lines (the positions of these sketches are estimated, the rings were centred around the middle of cylinders). These domain boundaries seem to cut the blurred broad interfaces at arbitrary positions, as no sharp interface is visible.

Thus, the identification of the interface position and domain size from STEM images is not obvious. One reasonable way to determine an average interface position is shown exemplarily for the lamellae-forming BCP. Figure 2d shows a line plot of the contrast cutting perpendicularly through three parallel lamellae of alternating PS and PMMA as in Figure 2c. The contrast distribution and thus the composition profile follows a sinusoidal curve (red fit). This is in good agreement with resonant soft X-ray reflectivity (RSoXR) observations by Sunday et al. [21] who found this sinusoidal material distribution being specific for BCPs with a comparably low χ and $\chi N \lesssim 23$, as apparent in PS-*b*-PMMA BCPs (Table 1). Figure 2d shows, that no plateau is formed in the composition profile, thus, no regime of completely separated pure phases is apparent. Thus, an experimental determination of the width of the domain interface is not feasible. Theoretical values of the interfacial width were determined following Equation (4) (Table 1) to 4.4–4.5 nm and are marked in the STEM-ADF images in Figure 2 for illustration (white solid lines). In literature, the width of PS-PMMA interfaces in similar BCPs is determined by neutron scattering to 5 nm [22]. If one considers the position of the domain interface at 50% concentration of one or the other polymer species, i.e., a locally predominant composition of either

PS or PMMA, it is possible to estimate the position of the interface. The resulting average domain sizes designating the 50% concentration threshold are listed in Table 1 and marked in Figure 2a–c by yellow dotted lines. The resulting domain sizes differ from those determined by SEM showing that SEM underestimates the PS domain sizes.

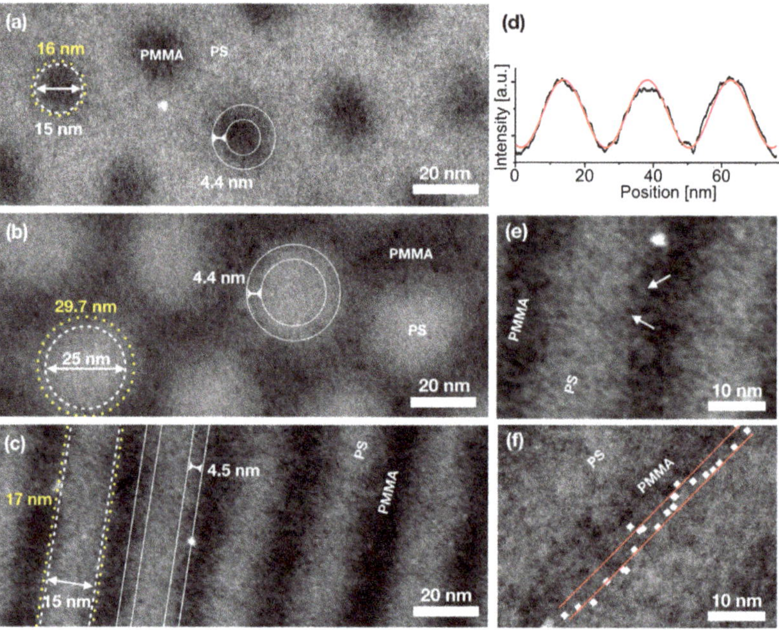

Figure 2. STEM-annular dark field (ADF) images of PS-*b*-PMMA BCPs with block length ratios of (**a**) PS:PMMA 70:30, (**b**) PS:PMMA 30:70 and (**c**) PS:PMMA 50:50 with bright contrast of PS and darker contrast of PMMA. Sketches in these images mark the domain sizes as measured by SEM (white dashed lines), domain sizes determined at 50% contrast in STEM ADF (yellow dotted lines) and theoretical interfacial widths of 4.4 and 4.5 nm (white solid lines). (**d**) Contrast profile perpendicular to alternating PS and PMMA lamellae as in (**c**) (black) superimposed by a sinusoidal fit (red). (**e**) Image section of BCP PS:PMMA 50:50 at higher magnification. White arrows mark grains of opposite contrast which might contain entrapped foreign polymer. (**f**) Image section of BCP PS:PMMA 50:50 with identified positions of 50% contrast (white dots) and resulting line edge roughness (LER) (distance of red lines = 3.2 nm).

The high-magnification STEM images show an additional internal structure of the polymers indicating strong compositional fluctuations along the lamellae as well as within or around cylinders. This internal structure is not accessible with SEM, AFM or other techniques. These fluctuations most likely result from the interfacial width and positional fluctuations of the domain interface along the long-axis of the lamella as well as along their through-film dimension. It was shown by Segal-Peretz et al. [58,59] using TEM tomography on stained PS-*b*-PMMA cylinders that the domain morphology through the polymer film is strongly distorted compared to a perfect cylinder. In all (S)TEM images shown in this work one analyses the projection of all these positional fluctuations.

One more striking observation is the existence of sharply bordered grains of opposite contrast within domains close to their 50%-interface. Such grains are shown in Figure 2e in higher magnification and some are exemplarily marked by arrows. As contrasts are assigned to different materials and as these grains appear in both polymer domains, it is likely that they contain the opposite material trapped within the foreign domain. Such nanoscale domains could form when polymer chains are trapped with contrariwise chain orientation not able to overcome the energy barrier to reorient during microphase separation, or with polymer chains being fully incorporated with both ends inside one polymer domain. In order to minimise interfacial areas, the entrapped parts of chains will then form polymer coils. This hypothesis is further supported by the observation that the frequency is much lower at the polymer domain interfaces of cylinder-forming BCPs than at those of the lamellae-forming BCP. From a geometrical point of view, the release of polymer ends within the wrong polymer domain, i.e., the healing of such defects, is favoured at curved interfaces compared to planar interfaces as the likelihood for the chains to reach an interface in close proximity is larger at curved than planar interfaces. It is also possible that the grains are agglomerates of random copolymer chains which were released from the functionalized SiO_2 substrate during polymer membrane preparation by HF etching. Such patches could adhere to the BCP film or be redeposited during the skimming of the polymer membranes with the TEM grid. However, this interpretation cannot convincingly explain the predominantly inverted contrast of these grains compared to their surroundings.

All these observed features at the polymer domain interfaces (interface width, positional fluctuations, grains) are expected to contribute to in the line edge roughness (LER) of domains. The LER was determined analysing the positional fluctuations of the 50% concentration threshold at several positions along a domain boundary. If one identifies the midpoint between absolute minima and maxima in several line scans, as in Figure 2d, along one domain one can determine the positional interface fluctuation, i.e., the LER. In case of the lamellae (PS:PMMA 50:50), the LER amounts to 1.66 ± 0.46 nm. This value is of the same order of magnitude as the theoretically expected value of 0.93 nm (Table 1) determined using Equation (5). The larger value can be explained by the superposition of interface fluctuations not only along one lamella but also along the domain interfaces through the polymer film. As stated above, it was observed in Reference [59] that the polymer domains are not perfectly cylindrically shaped but their morphology is irregular. It was also shown that the fluctuations of the interface position become even stronger towards the substrate and towards the interface with air/vacuum than they are within the polymer film. As our investigations are based on the projection of the domain interfaces throughout the film, a larger LER is expected. However, an LER of 1.66 nm can typically not be found on nanopatterns created using similar BCP lamellae as lithography template. LER of replicated patterns are usually much larger and measure approximately 4.8 nm [64]. It is likely that the larger LER results from the grains found close to the domain interfaces (Figure 2e). One can include these grains into the determination of the LER by not measuring the midpoint between absolute minima and maxima in the line plot (which neglects the existence of these grains), but defining the transition from >50% to <50% intensity as interface position. These positions are marked by white dots in Figure 2f. If one includes these grains in this way, the positional fluctuation doubles and measures 3.2 ± 1.9 nm. This observation also supports that the grains are no artefacts from sample preparation but features within the microphase separated polymer film.

In case of cylinder-forming BCPs (PS:PMMA 70:30 and 30:70), the large grains occur to a smaller extend, i.e., they poorly influence the line edge roughness of domains. The LER of these cylinder-forming BCPs can be determined to 1.78 nm for the BCP PS:PMMA 70:30 in Figure 2a and to 1.29 nm for the BCP PS:PMMA 30:70 in Figure 2b. Again, it is assumed that the curvature of the domains allows for the formation of smoother domain interfaces.

3.2. Spatially Resolved Investigation of the Composition of the Polymer Film Using (S)TEM-EELS

Electron energy loss spectra (EELS) were measured in TEM mode as well as STEM mode to investigate the chemical composition of assigned PS and PMMA domains. Spectra were collected using dual-EELS detection allowing for high integration times as well as precise determination of energies.

Combined EEL spectra of PS and PMMA taken from the 50:50 PS:PMMA BCP in the low-loss region and the high-loss region are shown in Figure 3a,b, respectively. These spectra were taken in TEM mode at 60 kV, the zero loss peak having a FWHM of 0.65 eV. In the low-loss region (Figure 3a), two distinct peaks are detected at 7 eV and at approximately 21 eV energy loss. The peak at 7 eV can be assigned to a π–π^* excitation of electrons in the aromatic ring of PS [65,66] verifying the existence of intact PS. The bulk plasmon peaks of PS and PMMA are found at 22 eV and 20 eV, respectively [67,68], appearing as a broad peak around 21 eV loss in this spectrum. Figure 3b shows the high-loss region of the EEL spectrum at the carbon K-edge. The near-edge fine structure of the C-K edge allows to differentiate between carbon binding states. In particular, the distinct peak at 285 eV loss corresponding to a C1s→π* (C=C) transition [65] again provides evidence of the presence of polystyrene, since a C=C bond does not exist in PMMA and thus PMMA does not show such a peak [67], while C-H bonds and C-C bonds can be found in both, PS and PMMA.

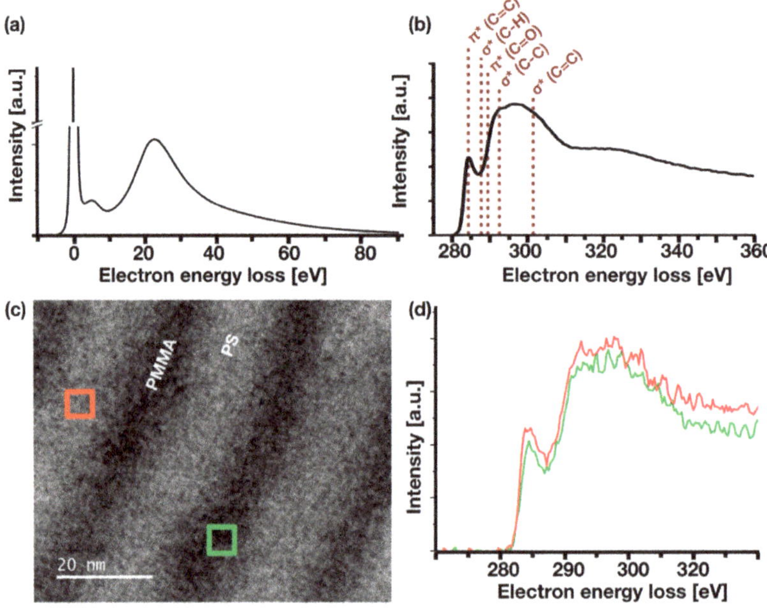

Figure 3. TEM-electron energy loss (EEL) spectra of BCP PS:PMMA 50:50 showing (**a**) the low-loss region and (**b**) high-loss at the carbon K-edge with assignment of peaks in (**b**) according to References [65–68]. (**c**) STEM-ADF image with marked positions of corresponding STEM-EELS spectra in (**d**). The spectra are displayed after zero-loss peak alignment and background subtraction.

STEM-EELS was employed to investigate the local distribution of PS and PMMA in the phase separated system. Figure 3c shows a STEM-ADF image with marked positions for the acquisition of the STEM-EEL spectra in Figure 3d. Spectra are very similar, which supports the hypothesis that no pure polymer domains are formed as already indicated by the sinusoidal contrast distribution found in STEM-ADF images (Figure 2d). The spectra in Figure 3d show clearly that PS exists at each sample position since the specific C1s→π* (C=C) transition at 285 eV can be found in both spectra of a PS domain and a PMMA domain. This verifies that no pure PMMA domains are present.

3.3. EFTEM Spectroscopic Imaging of PS and PMMA Lamellae

EFTEM spectroscopic imaging was performed to identify energy windows particularly suitable for the investigation of the PS and PMMA domains. Figure 4 displays a tableau of images collected using energy windows between −5 eV and +115 eV. Each image was collected using a filter slit width of 10 eV; the energy centre is noted in each image. While at higher electron energies typically smaller slit widths are used in EFTEM-SI, at 60 keV the slit width of 10 eV allowed for images without visible distortions. Two energy windows were found to be particularly suitable for high-resolution and high-contrast imaging of the PS and PMMA domains:

A.   **0 eV.** EFTEM zero loss imaging is known to increase contrasts in copolymers by removing inelastically scattered electrons from the image. This technique was introduced by Kunz et al. [69] and applied to copolymer blends [44]. Compared to conventional bright-field TEM images as shown in Figure 1h, the contrast in this image is increased and the resolution is high allowing for imaging of the internal structure of the polymer domains. This is also clearly visible in the zero loss images of both cylinder-forming BCPs (Figures S2 and S3). It is to note that an inversion of the contrasts between PS and PMMA in this zero-loss region occurs: PMMA appears brighter in TEM bright-field and zero-loss filtered EFTEM images than PS. In energy filtered images with the energy window centred at values between 10 and 110 eV PS rich domains appear brighter than their PMMA rich surroundings. This contrast inversion most likely results from a comparatively large plasmon excitation in PS compared to PMMA, since the low-loss region between 10 and 110 eV is dominated by the plasmon peak as visible in Figure 3a.

B.   **20–60 eV.** This energy region around and above the plasmon resonances of PS and PMMA [66–68] gives the best material contrasts. The internal structure of polymer domains with grains of inverse contrast, as discussed on the STEM-ADF images in Figure 2, become particularly visible in images obtained at 25–44 eV. The presence of 'bright' grains in a darker PS surrounding can be interpreted as the presence of PMMA inclusions leading to a locally enhanced plasmonic energy loss. Vice versa, the presence of 'darker' grains in an environment of bright PMMA rich surrounding would indicate a lack of PMMA material due to the inclusion of small PS grains. However, it is not possible to exclude that such 'inclusions' are actually located at the surface and are residuals of the random copolymer brush layer. In any case, it is likely that these grains contain the opposite polymer species.

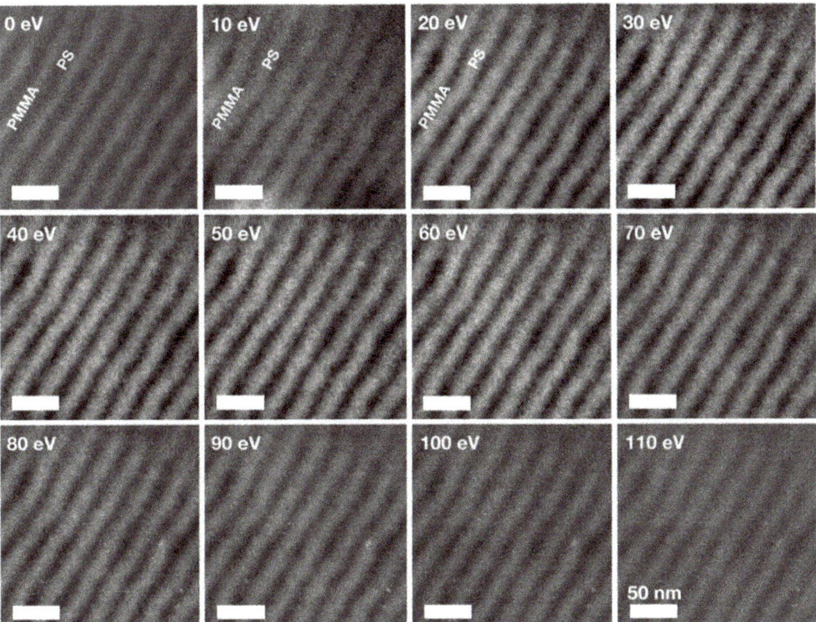

Figure 4. Energy-filtered TEM spectroscopic imaging of lamellae-forming BCP PS:PMMA 50:50. Images are collected from energy windows between −5 eV and +115 eV with an energy filter slit width of 10 eV. Energies noted in each image refer to the window centre. All scale bars are 50 nm.

Between the two regimes highlighted above a point of contrast inversion is found around 10 eV. Here, the contrast between polymer domains is poor and the PS domains appear very broad. At higher energies from 65 eV up to 115 eV contrasts are vanishing.

A video showing a series of images through the whole energy spectrum allows for direct comparison of the contrasts and can be found in the Supplementary Information. Tableaus of images of the two cylinder-forming block copolymers PS:PMMA 70:30 and PS:PMMA 30:70 obtained at these energy windows can be found in the Supplementary Information Figures S2 and S3, respectively.

3.4. Determination of Material Density Distributions by EFTEM Thickness Mapping

Thickness maps determined by energy-filtered TEM (EFTEM) are shown in Figure 5 for the BCPs (a) PS:PMMA 70:30, (b) PS:PMMA 30:70 and (c) PS:PMMA 50:50. Images were acquired at 60 kV with an integration time of 1.2 s/frame. Maps of the thickness t were measured using the t/λ-method [70], i.e., comparison of intensities in an unfiltered image and a filtered image with a 10 eV slit width at zero energy loss allowing only elastically scattered electrons to be detected. Consequently, the maps show the thickness in units of projected mean free paths λ (mfp) of electrons through the polymer domains. Using the model by Iakoubovskii [61], one can estimate the mfp at a given material mass density. The densities of PS and PMMA are assumed to 1.052 g/cm^3 and 1.159 g/cm^3, respectively [71]. The mean free paths then translate to 85.6 nm in PMMA and 87.6 nm in PS, i.e., they differ by 2.3%.

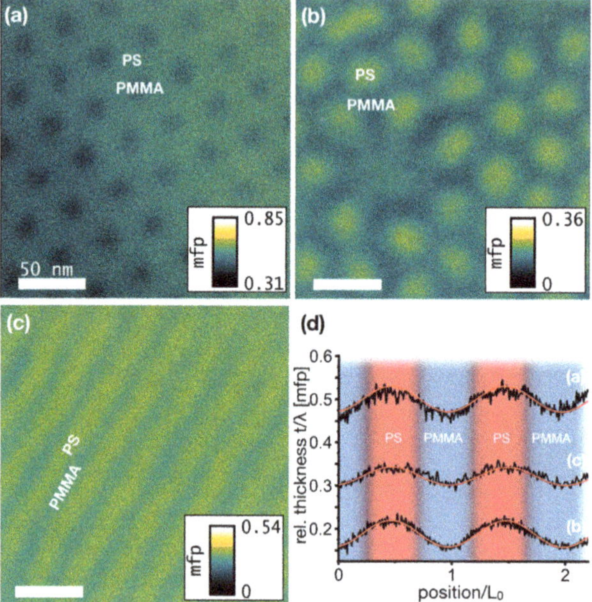

Figure 5. Energy-filtered TEM thickness maps of three BCPs with block length ratios of (**a**) PS:PMMA 70:30, (**b**) PS:PMMA 30:70 and (**c**) PS:PMMA 50:50. Colour coded maps display film thickness in units of mean free path (mfp). Lateral scale bars are all 50 nm. The small gradient in (**a**) from the lower left to the upper right corner is due to bending of the free polymer membrane close to a hole in the supporting Quantifoil film. (**d**) Line plots across polymer domains (position normalised to L_0 of each polymer).

In all three polymers, the PS domains exhibit a larger thickness than the PMMA domains (Figure 5). Thus, for all three polymers thickness profiles across polymer domains show a periodical behaviour (Figure 5d). For better comparison, the position axis is given in units of L_0 and curves are shifted such that the maxima match their positions. The local variation of t/λ along polymer domains follows a sinusoidal trend as was determined from contrast evaluation in STEM-ADF images in Figure 2d. If one takes into account that the mean free path of electrons is slightly larger in PS than in PMMA it becomes obvious that the thickness oscillations of block copolymer membranes are even more pronounced than those of t/λ. The average film thicknesses differ for the three polymers and amount to 43, 17 and 28 nm for the 70:30, 30:70 and 50:50 PS:PMMA films, compared to the targeted thicknesses of 35, 35 and 40 nm, respectively. Height differences between PS and PMMA domains are similar in both cylinder-forming BCPs, measuring 5.2 ± 0.2 nm. In case of the lamellae-forming BCP the height difference is with 3.5 nm significantly smaller, which might result from its smaller χN compared to the cylinder-forming BCPs (see Table 1) leading to a less efficient phase separation, thus, a stronger interpenetration of polymer chains into the opposite domain, as well as a larger density of entrapped grains of opposite polymer species (as shown in Figure 2e).

The local height differences between PS and PMMA domains appear inverted to those measured by AFM, where PMMA exceeds the PS level. It was, however, shown by Pérez-Murano and coworkers [72] that the larger elastic modulus of PMMA compared to PS can lead to such measurement artefacts during tapping mode AFM analysis of thin polymer films. It is also to note that the mass densities ρ used for the calculation of absolute polymer thicknesses are determined from bulk polymers. A different material density in microphase separated domains compared to bulk material could be apparent due to geometrical considerations of polymer chain configuration. According to the Iakoubovskii model, the mean free path λ depends on the mass density ρ as $\lambda \sim \rho^{0.3}$, and therefore mass densities ρ influence the conversion of measured maps to thickness maps.

Thus, assuming that the polymer film thickness is known, the EFTEM t/λ mapping method allows to visualise the density distribution within the phase separated polymer film and therefore gives further insight into the polymer mixing behaviour.

The PS matrix of the PS:PMMA 70:30 in Figure 5a, for instance, reveals a homogeneous t/λ distribution without any local variations. Due to geometrical considerations one could expect PS chains being less dense at triple points between three PMMA cylinders compared to the area between two neighbouring PMMA cylinders. This is, however, not found here, i.e., polymer chains most likely stretch, bend and compress to form a homogeneous PS matrix with minimal density fluctuations of less than 3.5%.

The image section of a PS:PMMA 30:70 BCP in Figure 5b shows a defect in the PS cylinder assembly: One PS cylinder is missing in the lower left part. This missing cylinder has a 7-fold coordinated environment to its nearest neighbours. The map reveals a slightly larger relative thickness t/λ at the expected position of the missing cylinder than in the surrounding PMMA matrix. This refers to a larger material density and/or thicker film at this exact position than in the PMMA matrix, indicating a larger PS concentration than in the surrounding PMMA matrix. Additionally, regions of smaller t/λ connect this spot to the neighbouring PS cylinders. Again, these connection lines are likely to contain increased PS concentrations. This polymer distribution suggests an insufficient microphase separation and a defect which is trapped in a metastable configuration within the process of forming distinct polymer domains.

3.5. Fraction of Non-Separated Polymers

The projected amount of any elemental species (including carbon) present at each position in a sample can be quantitatively determined using the EFTEM three-windows technique [73]. In this technique, an elemental map is calculated from three images, of which two are taken at different energies below a characteristic energy edge (for background calculation) and one in an energy window above the characteristic energy edge. After subtracting the extrapolated background from the post-edge image one obtains an image where the intensity is proportional to the number of atoms present in the sample integrated over the specimen thickness. Figure 6a displays such an EFTEM carbon map of the lamellae-forming BCP PS:PMMA 50:50 using the characteristic carbon K edge. Figure 6b shows a linear carbon concentration profile measured perpendicularly to the polymer lamellae as indicated by the red box. Since the ratio of molar carbon concentrations in pure PS and PMMA is 2:1, one would assume the concentration profile to exhibit oscillations with amplitudes of the same ratio, or more precisely with a ratio of 2.27:1, if one takes the different thicknesses of PS and PMMA lamellae (see Section 3.4) into account. Obviously, the projected carbon concentration oscillates by a much smaller amount, which can be attributed to the intermixing between PS and PMMA. Assuming that the volume fractions φ of PS inclusions in PMMA domains and of PMMA inclusions in PS domains are identical for the 50:50 BCP, one can calculate a volume fraction of $\varphi = 20\%$ of inclusions of the opposite polymer in the centre of each polymer lamella. Details of the calculation are given in the Supplementary Information.

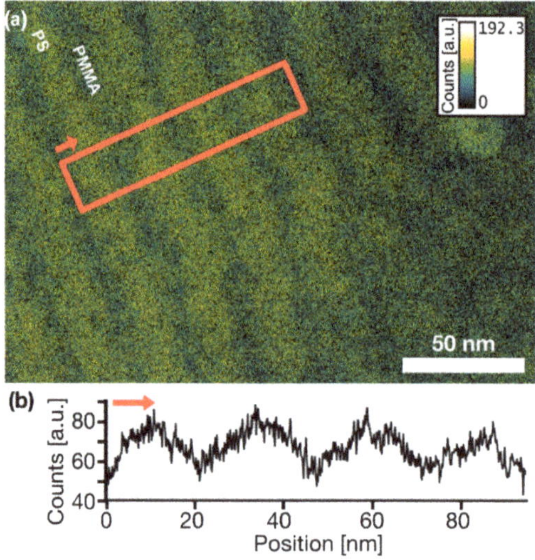

Figure 6. (**a**) Energy filtered TEM (EFTEM) carbon map and (**b**) concentration line-plot corresponding to the red marked area. The arrows display the direction of the line-plot.

4. Discussion

Self-assembled block copolymer nanostructures, which are widely used as templates for next generation nanolithography, were analysed by analytical (scanning) transmission electron microscopy ((S)TEM). In this work, cylindrical and lamellar nanodomains of PS and PMMA in microphase separated BCP thin films are imaged without staining at highest resolution using electrons at an energy as low as 60 keV. In contrast to more commonly used reciprocal space methods, real space imaging using analytical (S)TEM allows to correlate the internal structure of the polymer domains with characteristic parameters of these polymer patterns such as domain size, interface width and line edge roughness.

In particular, STEM dark-field images are presented, which reveal the internal structure of PS and PMMA domains. A sinusoidal contrast distribution along polymer domains allows to conclude that no pure domains containing one single polymer species are present but that periodical concentration gradients form the assigned PS or PMMA domains. This poor polymer separation might be due to the small Flory-Huggins parameter of PS and PMMA as well as the presence of a thick random copolymer brush locally promoting [74] and appearing like polymer intermixing. Thus, the term of an 'interfacial width' must be used carefully for the popular PS-*b*-PMMA BCPs. Line edge roughnesses (LER) were determined estimating the positional fluctuations of an interface at 50% polymer composition. One might assume that the LER measured in the projections of STEM or TEM images are affected by a possible tilting or wiggling of domain boundaries through the film. However, the isotropic smearing of interfaces around cylindrical domains in all directions suggests that such wiggling occurs only on a molecular level. Domain wiggling is observable in lamellar BCP films in in-plane directions, but on a wavelength which is large compared to the BCP film thickness. Therefore, such wiggling should not add very much to the LER observed. It is shown, however, that polymer grains, found particularly frequently in the lamellae-forming polymer, contribute largely to the line edge roughness of nanodomains. Electron energy loss spectroscopy (EELS) and energy filtered TEM (EFTEM) spectroscopic imaging were applied to investigate the chemical composition of the polymer domains. Acquiring images of polymer domains using electrons with a particular energy loss close to the bulk plasmon peak, allows for high-resolution imaging of the unstained polymers with strong material contrasts. EFTEM thickness maps were acquired revealing the density and film thickness distribution of phase separated BCPs giving insights into the spatial polymer distribution e.g., at defects in the polymer pattern. For the lamellae-forming BCP it is shown that the degree of microphase separation can be determined by analytical (S)TEM. It is found that even for the long-term annealing conditions applied here, a minimum of 20% volume fraction of non-separated polymer species is contained in the microphase separated lamellae, while in the case of cylinder-forming BCPs the fraction might be lower due to geometrical advantages.

Our analytical (S)TEM investigations shed light on the internal structure of polymer domains and polymer domain morphology, which impact pattern replication for lithography and infiltration purposes directly. These insights help elucidate the origin of line edge roughnesses of replicated nanopatterns and the limits in accuracy of selective infiltration as well as transport mechanisms in functional BCPs. It will be interesting to apply analytical STEM to investigate the properties of homopolymer blends or BCP-homopolymer blends which are often used to improve pattern order. Promising findings can also be expected when investigating other BCP species than PS-*b*-PMMA, such as high-χ polymers used to form sub-10 nm pitch patterns, where polymer segregation is more efficient and thus pure polymer domains with a smaller interfacial width are expected. Such high-χ polymer often include one Si-containing polymer species, thus, it can be anticipated that imaging with the above shown techniques will be even more facile.

Supplementary Materials: The following are available online at http://www.mdpi.com/2079-4991/10/1/141/s1, Figure S1: Comparison of bright-field TEM at 200 kV and 60 kV acceleration voltage, Video S1: EFTEM-SI of PS-PMMA 50:50, Figure S2: EFTEM-SI image series of 70:30 PS:PMMA BCP., Figure S3: EFTEM-SI image series of 30:70 PS:PMMA BCP.

Author Contributions: Conceptualization, K.B.; methodology, J.B., J.K.N.L.; investigation, J.B., V.S.K., D.K., K.B., J.K.N.L.; writing—original draft preparation, K.B.; writing—review and editing, K.B., J.K.N.L.; visualization, J.B.; supervision, K.B. and J.K.N.L. All authors have read and agreed to the published version of the manuscript.

Funding: This research was partially funded by the state of Northrhine Westfalia via the Forschungskolleg 'Leicht-Effizient-Mobil' (LEM).

Acknowledgments: The authors thank Mirko Schaper, Dept. of Mechanical Engineering at Paderborn University, for providing access to the SEM.

Conflicts of Interest: The authors declare no conflict of interest. The funders had no role in the design of the study; in the collection, analyses, or interpretation of data; in the writing of the manuscript, or in the decision to publish the results.

References

1. Bates, C.M.; Bates, F.S. 50th anniversary perspective: Block polymers-pure potential. *Macromolecules* **2016**, *50*, 3–22. [CrossRef]
2. Luo, M.; Epps, T.H. Directed block copolymer thin film self-assembly: Emerging trends in nanopattern fabrication. *Macromolecules* **2013**, *46*, 7567–7579. [CrossRef]
3. Darling, S.B. Directing the self-assembly of block copolymers. *Prog. Polym. Sci.* **2007**, *32*, 1152–1204. [CrossRef]
4. Ferrarese Lupi, F.; Giammaria, T.J.; Miti, A.; Zuccheri, G.; Carignano, S.; Sparnacci, K.; Seguini, G.; De Leo, N.; Boarino, L.; Perego, M.; et al. Hierarchical order in dewetted block copolymer thin films on chemically patterned surfaces. *ACS Nano* **2018**, *12*, 7076–7085. [CrossRef]

5. Brassat, K.; Kool, D.; Bürger, J.; Lindner, J.K.N. Hierarchical nanopores by block copolymer lithography on surfaces of different materials pre-patterned by nanosphere lithography. *Nanoscale* **2018**, *10*, 10005–10017. [CrossRef]

6. Bates, F.S.; Fredrickson, G.H. Block copolymer thermodynamics: Theory and experiment. *Annu. Rev. Phys. Chem.* **1990**, *41*, 525–557. [CrossRef]

7. Leibler, L. Theory of microphase separation in block copolymers. *Macromolecules* **1980**, *13*, 1602–1617. [CrossRef]

8. Hu, H.; Gopinadhan, M.; Osuji, C.O. Directed self-assembly of block copolymers: A tutorial review of strategies for enabling nanotechnology with soft matter. *Soft Matter* **2014**, *10*, 3867–3889. [CrossRef]

9. Brassat, K.; Lindner, J.K.N. Nanoscale block copolymer self-assembly and microscale polymer film dewetting: Progress in understanding the role of interfacial energies in the formation of hierarchical nanostructures. *Adv. Mater. Interfaces* **2019**, 1901565. [CrossRef]

10. Brassat, K.; Kool, D.; Nallet, C.G.A.; Lindner, J.K.N. Understanding film thickness-dependent block copolymer self-assembly by controlled polymer dewetting on pre-patterned surfaces. *Adv. Mater. Interfaces* **2019**, 190165. (In press) [CrossRef]

11. Brassat, K.; Kool, D.; Lindner, J.K.N. Modification of block copolymer lithography mask by O2/Ar plasma treatment: Insights from lift off experiments, nanopore etching and free membranes. *Nanotechnology* **2019**, *30*, 225302. [CrossRef] [PubMed]

12. Zhang, Y.; Sargent, J.L.; Boudouris, B.W.; Phillip, W.A. Nanoporous membranes generated from self-assembled block polymer precursors: Quo vadis? *J. Appl. Polym. Sci.* **2015**, *132*, 41683.

13. Yu, H.; Qiu, X.; Moreno, N.; Ma, Z.; Calo, V.M.; Nunes, S.P.; Peinemann, K.V. Self-assembled asymmetric block copolymer membranes: Bridging the gap from ultra-to nanofiltration. *Angew. Chem. Int. Ed.* **2015**, *54*, 13937–13941. [CrossRef] [PubMed]

14. Takashi, I.; Ghimire, G. Electrochemical applications of microphase-separated block copolymer thin films. *ChemElectroChem* **2018**, *5*, 2937–2953.

15. Steinhaus, A.; Srivastva, D.; Nikoubashman, A.; Gröschel, A.H. Janus Nanostructures from ABC/B Triblock Terpolymer Blends. *Polymers* **2019**, *11*, 1107. [CrossRef] [PubMed]

16. Majewski, P.W.; Gopinadhan, M.; Osuji, C.O. The effects of magnetic field alignment on lithium ion transport in a polymer electrolyte membrane with lamellar morphology. *Polymers* **2019**, *11*, 887. [CrossRef]

17. Young, W.S.; Kuan, W.F.; Epps, T.H., III. Block copolymer electrolytes for rechargeable lithium batteries. *J. Polym. Sci. B Polym. Phys.* **2014**, *52*, 1–16. [CrossRef]

18. Kuang, H.; Janik, M.J.; Gomez, E.D. Quantifying the role of interfacial width on intermolecular charge recombination in block copolymer photovoltaics. *J. Polym. Sci. B Polym. Phys.* **2015**, *53*, 1224–1230. [CrossRef]

19. Cianci, E.; Nazzari, D.; Seguini, G.; Perego, M. Trimethylaluminum Diffusion in PMMA Thin Films during Sequential Infiltration Synthesis: In Situ Dynamic Spectroscopic Ellipsometric Investigation. *Adv. Mater. Interfaces* **2018**, *5*, 1801016. [CrossRef]

20. Peng, Q.; Tseng, Y.C.; Mane, A.U.; DiDona, S.; Darling, S.B.; Elam, J.W. Effect of nanostructured domains in self-assembled block copolymer films on sequential infiltration synthesis. *Langmuir* **2017**, *33*, 13214–13223. [CrossRef]

21. Sunday, D.F.; Maher, M.J.; Hannon, A.F.; Liman, C.D.; Tein, S.; Blachut, G.; Asano, Y.; Ellison, C.J.; Willson, C.G.; Kline, R.J. Characterizing the interface scaling of high X block copolymers near the order-disorder transition. *Macromolecules* **2017**, *51*, 173–180. [CrossRef] [PubMed]

22. Longanecker, M.; Modi, A.; Dobrynin, A.; Kim, S.; Yuan, G.; Jones, R.; Satija, S.; Bang, J.; Karim, A. Reduced Domain Size and interfacial width in fast ordering nanofilled block copolmyer films by direct immersion annealing. *Macromolecules* **2016**, *49*, 8563–8571. [CrossRef]

23. Sunday, D.F.; Kline, R.J. Reducing block copolymer interfacial width through polymer additives. *Macromolecules* **2015**, *48*, 679–686. [CrossRef]

24. Kim, J.M.; Hur, Y.H.; Jeong, J.W.; Nam, T.W.; Lee, J.H.; Jeon, K.; Kim, Y.; Jung, Y.S. Block copolymer with am extremely high block-to-block interaction for a significant reduction of line-edge fluctuations in self-assembled patterns. *Chem. Mater.* **2016**, *28*, 5680–5688. [CrossRef]

25. Helfand, E.; Tagami, Y. Theory of the interface between immiscible polymers. II. *J. Chem. Phys.* **1972**, *56*, 3592–3601. [CrossRef]

26. Broseta, D.; Fredrickson, G.H.; Helfand, E.; Leibler, L. Molecular weight and polydispersity effects at polymer-polymer interfaces. *Macromolecules* **1990**, *23*, 132–139. [CrossRef]

27. Helfand, E.; Tagami, Y. Theory of the interface between immiscible polymers. *J. Chem. Phys.* **1972**, *57*, 1812–1813. [CrossRef]

28. Helfand, E.; Wasserman, Z.R. Block copolymer theory. 4. Narrow interphase approximation. *Macromolecules* **1976**, *9*, 879–888. [CrossRef]

29. Shull, K.R. Mean-field theory of block copolymers: Bulk melts, surfaces, and thin films. *Macromolecules* **1992**, *25*, 2122–2133. [CrossRef]

30. Shull, K.R.; Mayes, A.M.; Russell, T.P. Segment distributions in lamellar diblock copolymers. *Macromolecules* **1993**, *26*, 3929–3936. [CrossRef]

31. Ermoshkin, A.V.; Semenov, A.N. Interfacial tension in binary polymer mixtures. *Macromolecules* **1996**, *29*, 6294–6300. [CrossRef]

32. Tang, H.; Freed, K.F. Interfacial studies of incompressible binary blends. *J. Chem. Phys.* **1991**, *94*, 6307–6322. [CrossRef]

33. Semenov, A.N. Theory of block copolymer interfaces in the strong segregation limit. *Macromolecules* **1993**, *26*, 6617–6621. [CrossRef]

34. Black, C.T.; Ruiz, R.; Breyta, G.; Cheng, J.Y.; Colburn, M.E.; Guarini, K.W.; Kim, H.-C.; Zhang, Y. Polymer self assembly in semiconductor microelectronics. *IBM J. Res. Dev.* **2007**, *51*, 605–633. [CrossRef]

35. Mocan, M.; Kamperman, M.; Leermakers, F.A.M. Microphase segregation of diblock copolymers studies by the self-consistent field theory of Scheutjens and Fleer. *Polymers* **2018**, *10*, 78. [CrossRef]

36. Foster, M.D.; Sikka, M.; Singh, N.; Bates, F.S.; Satija, S.K.; Majkrzak, C.F. Structure of symmetric polyolefin block copolymer thin films. *J. Chem. Phys.* **1992**, *96*, 8605–8615. [CrossRef]

37. Winey, K.L.; Thomas, E.L.; Fetters, L.J. Ordered morphologies in binary blends of diblock copolymer and homopolymer and characterization of their intermaterial dividing surfaces. *J. Chem. Phys.* **1991**, *95*, 9367–9375. [CrossRef]

38. Scherble, J.; Stark, B.; Stühn, B.; Kressler, J.; Budde, H.; Höring, S.; Schubert, D.W.; Simon, P.; Stamm, M. Comparison of interfacial width of block copolymers of d8-poly(methylmethacrylate) with various poly(n-alkylmethacrylate)s and the respective homopolymer pairs as measured by neutron reflection. *Macromolecules* **1999**, *32*, 1859–1864. [CrossRef]

39. Melekevitz, J.; Muthukumar, M. Density functional theory of lamellar ordering in diblock copolymers. *Macromolecules* **1991**, *24*, 4199–4205. [CrossRef]

40. Russell, T.P. On the reflectivity of polymers: Neutrons and X-rays. *Phys. B* **1996**, *221*, 267–283. [CrossRef]

41. Hannon, A.F.; Sunday, D.F.; Bowen, A.; Khaira, G.; Ren, J.; Nealey, P.F.; de Pablo, J.J.; Kline, R.J. Optimizing self-consistent field theory block copolymer models with X-ray metrology. *Mol. Syst. Des. Eng.* **2018**, *3*, 376–389. [CrossRef] [PubMed]

42. Anastasiadis, S.H.; Retsos, H.; Toprakcioglu, C.; Menelle, A.; Hadziioannou, G. On the interfacial width in triblock versus diblock copolymers: A neutron reflectivity investigation. *Macromolecules* **1998**, *31*, 6600–6604. [CrossRef]

43. Bates, F.S.; Schulz, M.F.; Khandpur, A.K.; Förster, S.; Rosedale, J.H.; Almdal, K.; Mortensen, K. Fluctuations, conformational asymmetry and block copolymer phase behaviour. *Faraday Discuss.* **1994**, *98*, 7–18. [CrossRef]

44. Gomez, E.D.; Ruegg, M.L.; Minor, A.M.; Kisielowski, C.; Downing, K.H.; Glaeser, R.M.; Balsara, N.P. Interfacial concentration profiles of rubbery polyolefin lamellae determined by quantitative electron microscopy. *Macromolecules* **2008**, *41*, 156–162. [CrossRef]

45. Majewski, P.W.; Yager, K.G. Rapid ordering of block copolymer thin films. *J. Phys. Condens. Matter* **2016**, *28*, 403002. [CrossRef]

46. Anastasiadis, S.H.; Russell, T.P.; Satija, S.K.; Majkrzak, C.F. The morphology of symmetric diblock copolymers as revealed by neutron reflectivity. *J. Chem. Phys.* **1990**, *92*, 5677–5691. [CrossRef]

47. Green, P.F.; Christensen, T.M.; Russell, T.P.; Jerome, R. Equilibrium surface composition of diblock copolymers. *J. Chem. Phys.* **1990**, *92*, 1478–1482. [CrossRef]

48. Russell, T.P.; Menelle, A.; Hamilton, W.A.; Smith, G.S.; Satija, S.K.; Majkrzak, C.F. Width of homopolymer interfaces in the presence of symmetric diblock copolymers. *Macromolecules* **1991**, *24*, 5721–5726. [CrossRef]

49. Anastasiadia, S.H.; Russell, T.P.; Satija, S.K.; Majkrzak, C.F. Neutron reflectivity studies of the surface-induced ordering of diblock copolymer films. *Phys. Rev. Lett.* **1989**, *62*, 1852. [CrossRef]

50. Stamm, M. Investigation of the interface between polymers: A comparison of scattering and reflectivity techniques. *J. Appl. Crystallogr.* **1991**, *24*, 651–658. [CrossRef]

51. Lee, J.; Kang, M.H.; Lim, W.C.; Shin, K.; Lee, Y. Characterization of microphase-separated diblock copolymer films by TOF-SIMS. *Surf. Interface Anal.* **2013**, *45*, 498–502. [CrossRef]

52. Wang, C.; Araki, T.; Ade, H. Soft X-ray resonant reflectivity of low-z material thin films. *Appl. Phys. Lett.* **2005**, *87*, 214109. [CrossRef]

53. Ade, H.; Wang, C.; Garcia, A.; Yan, H.; Sohn, K.E.; Hexemer, A.; Bazan, G.C.; Nguyen, T.Q.; Kramer, E.J. Characterization of multicomponent polymer trilayers with resonant soft X-ray reflectivity. *J. Polym. Sci. Part B Polym. Phys.* **2009**, *47*, 1291–1299. [CrossRef]

54. Yan, H.; Wang, C.; Garica, A.; Swaraj, S.; Gu, Z.; McNeill, C.R.; Schuettfort, T.; Sohn, K.E.; Kramer, E.J.; Bazan, G.C.; et al. Interfaces in organic devices studied with resonant soft X-ray reflectivity. *J. Appl. Phys.* **2011**, *110*, 102220. [CrossRef]

55. Jinnai, H.; Higuchi, T.; Zhuge, X.; Kumamoto, A.; Batenburg, K.J.; Ikuhara, Y. Three-dimensional visualization and characterization of polymeric self-assemblies by transmission electron microtomography. *Acc. Chem. Res.* **2017**, *50*, 1293–1302. [CrossRef]

56. Lohse, D.J.; Hadjichristidis, N. Microphase separation in block copolymers. *Curr. Opin. Colloid Interface Sci.* **1997**, *2*, 171–176. [CrossRef]

57. Kunz, M.; Shull, K. Improved technique for cross-sectional imaging of thin polymer films by transmission electron microscopy. *Polymer* **1993**, *34*, 2427–2430. [CrossRef]

58. Segal-Peretz, T.; Winterstein, J.; Doxastakis, M.; Ramirez-Hernandez, A.; Biswas, M.; Ren, J.; Suh, H.S.; Darling, S.B.; Liddle, J.A.; Elam, J.W.; et al. Characterizing the three-dimensional structure of block copolymers via sequential infiltration synthesis and scanning transmission electron tomography. *ACS Nano* **2015**, *9*, 5333–5347. [CrossRef]

59. Segal-Peretz, T.; Ren, J.; Xiong, S.; Khaira, G.; Bowen, A.; Ocola, L.E.; Divan, R.; Doxastakis, M.; Ferrier, N.J.; de Pablo, J.J.; et al. Quantitative three-dimensional characterization of block copolymer directed self-assembly on combined chemical and topographical prepatterned templates. *ACS Nano* **2017**, *11*, 1307–1319. [CrossRef]

60. Handlin, D.L.; Thomas, E.L. Phase contrast imaging of styrene-isoprene and styrene-butadiene block copolymers. *Macromolecules* **1983**, *16*, 1514–1525. [CrossRef]

61. Iakoubovskii, K.; Mitsuishi, K.; Nakayama, Y.; Furuya, K. Thickness measurements with electron energy loss spectroscopy. *Microsc. Res. Tech.* **2008**, *71*, 626–631. [CrossRef] [PubMed]

62. Russell, T.P.; Hjelm, R.P.; Seeger, P.A. Temperature dependence of the interaction parameter of polystyrene and poly (methyl methacrylate). *Macromolecules* **1990**, *23*, 890–893. [CrossRef]

63. Puglisi, R.A. Towards ordered silicon nanostructures through self-assembling mechanisms and processes. *J. Nanomater.* **2015**, *16*, 229. [CrossRef]

64. Lee, K.S.; Lee, J.; Kwak, J.; Moon, H.C.; Kim, J.K. Reduction of line edge roughness of polystyrene-block-polymethylmethacrylate copolymer nanopatterns by introducing hydrogen bonding at the junction point of two block chains. *ACS Appl. Mater. Interfaces* **2017**, *9*, 31245–31251. [CrossRef]

65. Yakovlev, S.; Libera, M. Dose-limited spectroscopic imaging of soft materials by low-loss EELS in the scanning transmission electron microscope. *Micron* **2008**, *39*, 734–740. [CrossRef]

66. Varlot, K.; Martin, J.M.; Quet, C. Physical and chemical changes in polystyrene during electron irradiation using EELS in the TEM: Contribution of the dielectric function. *J. Microsc.* **1998**, *191*, 187–194. [CrossRef]

67. Ritsko, J.J.; Brillson, L.J.; Bigelow, R.W.; Fabish, T.J. Electron energy loss spectroscopy and the optical properties of polymethylmethacrylate from 1 to 300 eV. *J. Chem. Phys.* **1978**, *69*, 3931–3939. [CrossRef]

68. Varlot, K.; Martin, J.M.; Quet, C. EELS analysis of PMMA at high spatial resolution. *Micron* **2001**, *32*, 371–378. [CrossRef]

69. Kunz, M.; Moller, M.; Cantow, H.J. The net distribution of elements by element specific electron microscopy-ESI, Morphological studies of polystyrene-block-poly (2-vinylpyridine). *Macromol. Chem. Rapid Commun.* **1987**, *8*, 401–410. [CrossRef]

70. Malis, T.; Cheng, S.C.; Egerton, R.F. EELS log ratio technique for specimen-thickness measurement in the TEM. *J. Electron Microsc. Tech.* **1988**, *8*, 193–200. [CrossRef]

71. Polymer Properties Database. Available online: http://polymerdatabase.com/polymer%20physics/Polymer%20Density.html (accessed on 29 November 2019).

72. Lorenzoni, M.; Evangelio, L.; Nicolet, C.; Navarro, C.; San Paulo, A.; Pérez-Murano, F. Nanomechanical properties of solvent cast PS and PMMA polymer blends and block copolymers. In Proceedings of the SPIE, San Jose, CA, USA, 27 March 2015; Volume 9423, p. 942325.

73. Egerton, R.F. *Electron Energy-Loss Spectroscopy in the Transmission Electron Micoscope*; Plenum Press: New York, NY, USA, 1996.

74. Sparnacci, K.; Antonioli, D.; Gianotti, V.; Laus, M.; Ferrarese Lupi, F.; Giammaria, T.J.; Seguini, G.; Perego, M. Ultrathin random copolymer-grafted layers for block copolymer self-assembly. *ACS Appl. Mater. Interfaces* **2015**, *7*, 1094–10951. [CrossRef] [PubMed]

 *nanomaterials*

Article

Tannic Acid-Mediated Aggregate Stabilization of Poly(*N*-vinylpyrrolidone)-*b*-poly(oligo (ethylene glycol) methyl ether methacrylate) Double Hydrophilic Block Copolymers

Noah Al Nakeeb [1], Ivo Nischang [2,3],* and Bernhard V.K.J. Schmidt [1],*

[1]   Max-Planck Institute of Colloids and Interfaces, Department of Colloid Chemistry, Am Mühlenberg 1, 14476 Potsdam, Germany; noah.alnakeeb@gmail.com

[2]   Laboratory of Organic and Macromolecular Chemistry (IOMC), Friedrich Schiller University Jena, Humboldtstraße 10, 07743 Jena, Germany

[3]   Jena Center for Soft Matter (JCSM), Friedrich Schiller University Jena, Philosophenweg 7, 07743 Jena, Germany

*   Correspondence: ivo.nischang@uni-jena.de (I.N.); bernhard.schmidt@mpikg.mpg.de (B.V.K.J.S.)

Received: 3 April 2019; Accepted: 24 April 2019; Published: 26 April 2019

Abstract: The self-assembly of block copolymers in aqueous solution is an important field in modern polymer science that has been extended to double hydrophilic block copolymers (DHBC) in recent years. In here, a significant improvement of the self-assembly process of DHBC in aqueous solution by utilizing a linear-brush macromolecular architecture is presented. The improved self-assembly behavior of poly(*N*-vinylpyrrolidone)-*b*-poly(oligo(ethylene glycol) methyl ether methacrylate) (PVP-*b*-P(OEGMA)) and its concentration dependency is investigated via dynamic light scattering (DLS) (apparent hydrodynamic radii ≈ 100–120 nm). Moreover, the DHBC assemblies can be non-covalently crosslinked with tannic acid via hydrogen bonding, which leads to the formation of small aggregates as well (apparent hydrodynamic radius ≈ 15 nm). Non-covalent crosslinking improves the self-assembly and stabilizes the aggregates upon dilution, reducing the concentration dependency of aggregate self-assembly. Additionally, the non-covalent aggregates can be disassembled in basic media. The presence of aggregates was studied via cryogenic scanning electron microscopy (cryo-SEM) and DLS before and after non-covalent crosslinking. Furthermore, analytical ultracentrifugation of the formed aggregate structures was performed, clearly showing the existence of polymer assemblies, particularly after non-covalent crosslinking. In summary, we report on the completely hydrophilic self-assembled structures in solution formed from fully biocompatible building entities in water.

Keywords: block copolymer self-assembly; analytical ultracentrifugation; tannic acid

1. Introduction

Block copolymer self-assemblies play a prominent role in current polymer science [1,2]. Self-assemblies are applied in many fields of research such as (nano)-lithography [3,4], nanoparticle formation [5,6], and catalysis [7,8], but also in the biomedical field, where applications such as imaging [9,10], biological sensing [11,12], and drug delivery [13–15] are investigated. Amphiphilic block copolymers are utilized frequently for these tasks in an aqueous environment, e.g., via self-assembly to micelles, vesicles, or more complex structures [16,17]. Recently, the formation of amphiphilic self-assembled structures was shifted to the polymerization process [18,19]. In such a way, the aforementioned and more complex structures can be accessed in one step via the adjustment of monomer conversion. Notwithstanding, self-assemblies of amphiphilic block copolymers in aqueous solution face some disadvantages, e.g., low permeability [20]. Particularly, in the prospected applications of nanoreactors, permeability is a key property to enable efficient reaction progress. One solution to this problem is the introduction of artificial protein channels [21,22]. Another option is the implementation of stimuli-responsive blocks that allow selective swelling/deswelling of the hydrophobic domain [23,24].

Another way would be to switch from an amphiphilic block copolymer to a completely water-soluble block copolymer. Frequently, one of the utilized blocks is stimuli-responsive, which facilitates self-assembly. As the solubility changes after stimulus application in most cases, the self-assembled structures are not purely hydrophilic anymore. In the case of self-assembly based on the hydrophilic effect, no stimulus is employed to form self-assembled structures via a change in solubility. The self-assembly process of double hydrophilic block copolymers (DHBCs) in water can be related to the macroscopic two-phase formation of hydrophilic homopolymer mixtures [25], e.g., poly(ethylene glycol) (PEG) and dextran—two water soluble polymers that were utilized to purify proteins [26]—or in the formation of water-in-water emulsions [27,28]. As the blocks are connected covalently in the DHBC, no macroscopic demixing is possible; rather, a microphase separation occurs, which is termed the hydrophilic effect [29]. The process of demixing, and consequently weak aggregation, is related to the difference in hydrophilicity that leads to non-symmetric solvation. Therefore, a difference in osmotic pressure is present between the respective polymer domains, which has to be compensated via the aggregation of the polymer domains. Certainly, the architecture of the individual blocks has a significant effect as well due to interactions between the blocks and their structure in space, which can also be related to varying solvation. Recently, several examples of DHBC self-assembly were introduced, e.g., pullulan-*b*-poly(*N*,*N*-dimethylacrylamide) [30], poly(2-ethyl-2-oxazoline)-*b*-poly(*N*-vinylpyrrolidone) (PEtOx-*b*-PVP) [31], PEG-*b*-glycopolymer [32], poly(oligo(ethylene glycol) methyl ether methacrylate)-*b*-glycopolymer (POEGMA-*b*-glycopolymer) [33,34] or PEtOx-*b*-PEG [35]. Moreover, the effect of DHBC architecture was investigated, and it could be shown that a linear-brush DHBC showed significantly enhanced self-assembly behavior, i.e., the system pullulan-*b*-P(OEGMA) [36]. In addition, DHBCs were utilized in the formation of inorganic and metal–organic mesocrystals [37,38]. Considering the possible future application of DHBC in the biomedical field, a focus is on the utilization of biocompatible polymers, namely PEG and (PVP as a possible combination of particular interest [20,39]. For example, PVP is frequently used in drug formulations [40]. However, the self-assembly of linear PEG-*b*-PVP showed rather limited success [41]. Therefore, exchanging PEG with P(OEGMA) is a useful alternative, as recently shown by our group [42]. Nevertheless, one of the major challenges for DHBC self-assembly is the highly dynamic and thus poor stability of aggregates, especially under conditions of dilution. The stability of DHBC aggregates can be expressed with an equilibrium-like state between assembled aggregates and unimers in aqueous solution. The equilibrium strongly depends on concentration, and thus disassembly is observed in diluted solution. An option to circumvent the dynamics of DHBC self-assembly is the crosslinking of the formed structures at higher concentration, which renders the aggregates stable under dilution [41]. So far, covalent crosslinking was the focus of research, which essentially freezes the dynamics of the DHBC self-assembly system without the option of disassembly. A triggered disassembly could be achieved via dynamic covalent chemistry based on disulfide or imine bonds [43]. To obtain a true adaptive system, non-covalent crosslinking chemistries have to be introduced, i.e., supramolecular bonding, for example via hydrogen bonds.

The concept of supramolecular chemistry allows the introduction of dynamics into molecular systems [44,45], e.g., via hydrogen bonding [46], host–guest complexes [47], or metal complexation [48]. Supramolecular interactions are particularly interesting for reversible crosslinking [49,50]. In that regard, tannic acid (TA) is a very useful compound to non-covalently crosslink the formed self-assemblies from PVP-*b*-P(OEGMA), as the interaction of tannic acid and PVP is well known not only from science, but also from the fining of red wine. It readily forms hydrogen bonds due to its acidic phenolic hydroxyls with a significant number of molecules, such as proteins and polymers, including PVP [51], which can be exploited in the formation of nanostructures [52–54]. Moreover, TA belongs to the group of tannins and can be found in wine, beer, tea, and nuts. Therefore, TA represents a natural and renewable product that is also approved by the Food and Drug Administration. Therefore, TA gained a lot of attention recently, last but not least because of its utilization in biomedical applications [55,56].

Herein, the self-assembly of PVP-*b*-P(OEGMA) and crosslinking via biocompatible TA in aqueous solution is described (Scheme 1). The DHBC is synthesized via a combination of reversible deactivation radical polymerization and copper (I) catalyzed azide alkyne cycloaddition (CuAAc). Subsequently, dynamic light scattering (DLS), cryogenic scanning electron microscopy (cryo-SEM), and analytical ultracentrifugation (AUC) are utilized to investigate the desired aggregate formation. The selection of these techniques was based on the different insights provided based on the physical principle of their operation. Therefore, self-assembly efficiency and aggregate sizes are studied as well as the dynamics of the formed aggregates before and after non-covalent crosslinking.

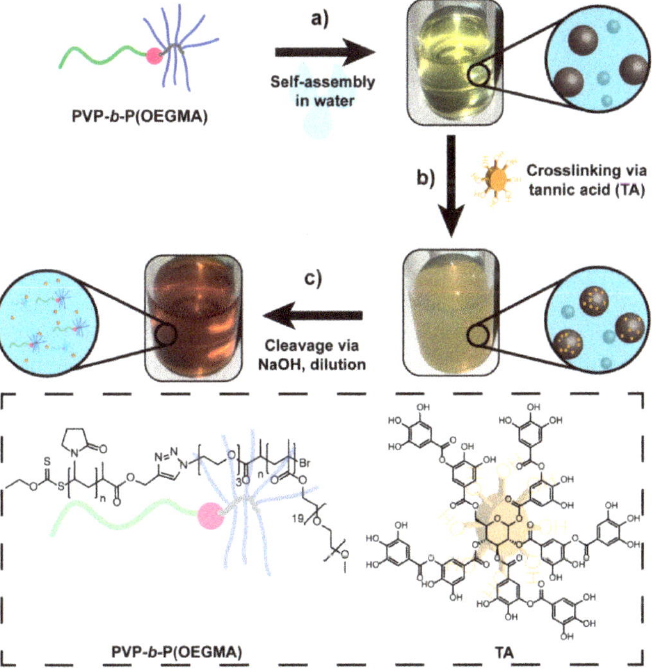

Scheme 1. Overview of the concept utilized in the present work. (**a**) The self-assembly process of the linear-brush double hydrophilic block copolymers (DHBC) poly(*N*-vinylpyrrolidone)-*b*-poly(oligo(ethylene glycol) methyl ether methacrylate) (PVP-*b*-P(OEGMA)) in water, followed by (**b**) crosslinking of the poly(*N*-vinylpyrrolidone) (PVP) blocks via tannic acid (TA), and finally (**c**) the disassembly process via the addition of NaOH. Inset: Structures of PVP-*b*-P(OEGMA) and TA (idealized).

2. Materials and Methods

Methods: ^{1}H- and ^{13}C-NMR spectra were recorded at ambient temperature at 400 MHz for ^{1}H and 100 MHz for ^{13}C with a Bruker Ascend400 (Bruker, Billerica, MA, USA). Dynamic light scattering (DLS) was performed using an ALV-7004 Multiple Tau Digital Correlator (ALV, Langen, Germany) in combination with a CGS-3 Compact Goniometer (ALV, Langen, Germany) and a HeNe laser (Polytec, 34 mW, $\lambda = 633$ nm at $\theta = 90°$ setup for DLS). Sample temperatures were adjusted to 25 °C and toluene was used as the immersion liquid. Apparent hydrodynamic radii (R_{app}) were determined by LV-Correlator Software Version 3.0. Autocorrelation functions measured at a scattering angle of 90° were analyzed using Constrained Regularization Method for Inverting Data (CONTIN). Cryogenic scanning electronic microscopy (cryo-SEM) was performed on a Jeol JSM 7500 F (Tokio, Japan) and the Alto 2500 cryo-chamber from Gatan (Munich, Germany). Size exclusion chromatography (SEC) for PVP, P(OEGMA), and PVP-*b*-P(OEGMA) were conducted in *N*-methyl pyrrolidone (NMP) (99%, Carl Roth) with 0.05 mol L^{-1} of LiBr and methyl benzoate as internal standard at 70 °C using a column system with a GRAM 100/1000 column (8 × 300 mm, 7-μ particle size) from PSS (Mainz, Germany), a GRAM precolumn (8 × 50 mm) from PSS, a Shodex RI-71 detector, and a poly(methyl methacrylate) (PMMA) calibration with standards from PSS. Fourier transform infrared (FT-IR) spectra were acquired on a Nicolet iS 5 FT-IR spectrometer (Thermo Fisher Scientific, Schwerte, Germany).

Sedimentation velocity experiments were performed with a ProteomeLab XL-I analytical ultracentrifuge (Beckman Coulter Instruments, Brea, CA, USA), using double-sector epon centerpieces with a 12-mm optical path length. The cells were placed in an An-50 Ti eight-hole rotor. A rotor speed of 42,000 rpm was used. The cells were filled with 420 μL of sample solution and with 440 μL of water as the reference. The experiments were conducted for 24 h at a temperature of T = 20 °C. Sedimentation profile scans were recorded with the interference optics (refractive index (RI)) detection system with respect to time at 2-min intervals. A suitable selection of scans was used for data evaluation with Sedfit using the ls-g*(s) model [57], i.e. by least squares boundary modeling with the implemented Tikhonov-Phillips regularization procedure and by assuming non-diffusing species. This model results in an apparent differential distribution of sedimentation coefficients, *s*.

Materials: Ammonium chloride (99%, Carl Roth, Karlsruhe, Germany), ascorbic acid (98%, Alfa Aesar, Karlsruhe, Germany), 2-bromopropionyl bromide (97%, Sigma Aldrich, Steinheim, Germany), *t*-butyl hydroperoxide (70% solution in water, Acros Organics, Geel, Belgium), chloromethyl polystyrene resin (2.4 mmol g^{-1}, TCI, Eschborn, Germany), copper (I) bromide (CuBr, 99.99%, Sigma Aldrich), copper (II) sulfate (CuSO4, 99%, Carl Roth), dichloromethane (DCM, analytical grade, Acros Organics), diethyl ether (ACS reagent, Sigma Aldrich), *N*,*N*-dimethylformamide (DMF, analytical grade, Sigma Aldrich), dimethylsulfoxide (DMSO, analytical grade, VWR Chemicals, Darmstadt, Germany), 4,4′-dinonyl-2,2′-dipyridyl (dNBipy, 97%, Sigma Aldrich), ethyl acetate (EtOAc, analytical grade, Chem Solute, Renningen, Germany), hexane (analytical grade, Fluka, Schwerte, Germany), hydrochloric acid (fuming, Carl Roth), magnesium sulfate (dried, Fisher Scientific, Schwerte, Germany), methanol (MeOH, analytical grade, Fisher Scientific), *N*,*N*,*N*′,*N*″,*N*″-pentamethyldiethylenetriamine (PMDETA, 98%, Sigma Aldrich), potassium-O-ethyl xanthate (98%, Alfa Aesar), propargyl alcohol (99%, Sigma Aldrich), pyridine (99% extra dry, Acros Organics), sodium hydroxide (NaOH, 98%, Sigma Aldrich), sodium azide (>99.5%, Fluka), sodium bicarbonate (>99%, Fluka), sodium sulfite (97%, Acros Organics), tannic acid (Alfa Aesar), tetrahydrofuran (THF, analytical grade, Fisher Scientific), and triethylamine (99.5%, Sigma Aldrich) were used as received. *N*-Vinylpyrrolidone (VP, 99%, Sigma Aldrich) was dried over anhydrous magnesium sulfate and purified by distillation under reduced pressure. Oligo(ethylene glycol) methyl ether methacrylate (OEGMA, 900 g mol^{-1}, Sigma Aldrich) was first dissolved in THF, and then passed over a basic aluminum oxide column (Brockmann I, Sigma Aldrich) and subsequently precipitated in cold hexane, filtered, and dried under high vacuum for 24 h. Millipore water was obtained from an Integra UV plus pure water system by SG Water (Hamburg, Germany). Azido functionalized poly(styrene)-resin, prop-2-yn-1-yl 2-((ethoxycarbonothioyl)thio) propanoate (alkyne-CTA), alkyne end-functionalized PVP (Figure S1), and azide end functionalized P(OEGMA) (Figure S2) were prepared according to the literature [30,36,58]. Spectra/Por dialysis tubes with a molecular weight cut-off of 10,000 were purchased from Spectrum Labs (Los Angeles, CA, USA).

Procedure for the preparation of DHBC aqueous solutions for DLS investigation: For the DLS measurements, 50.0 mg of DHBC were dissolved in 2.5 g of Millipore water in order to obtain a 2.0 wt.% DHBC solution. Before conducting the DLS measurement, the solution was passed through a 1.2 μm filter. The solutions containing 0.5 wt.% and 0.1 wt.% of DHBC were obtained by diluting the initial 2.0 wt.% DHBC solution with Millipore water accordingly.

Crosslinking of PVP-b-P(OEGMA) via TA: In order to crosslink the PVP-*b*-P(OEGMA) aggregates, a 0.5 wt.% TA solution was added to a 10.0 wt.% PVP-*b*-P(OEGMA) solution to form a 2.0 wt.%

PVP-*b*-P(OEGMA)-TA-solution. Then, the mixture of TA with the polymer was characterized via DLS, cryo-SEM, and AUC. The 0.5 wt.% and 0.1 wt.% solutions were obtained by diluting the 2.0 wt.% solutions accordingly.

Disassembly of PVP-b-P(OEGMA) TA crosslinking via NaOH: In order to induce a disassembly of the PVP-*b*-P(OEGMA) aggregates, first, a crosslinking reaction was applied as reported in previous sections receiving 2 g of a 2.0 wt.% crosslinked DHBC solution. Subsequently, 40 mg of NaOH was added to the solution to obtain a 0.5-M NaOH solution. The solution was stirred for 1 h before it was passed through a 1.2-μm filter followed by its investigation via DLS and cryo-SEM. The 0.5 wt.% and 0.1 wt.% solutions for the DLS investigations were obtained by diluting the 2.0 wt.% solution accordingly.

AUC measurements of PVP-b-P(OEGMA): For the AUC measurements, two stock solutions were used. The 2.0 wt.% stock solution of pure DHBC in Millipore water was diluted further to afford the following polymer concentrations: 2.0 wt.%, 1.0 wt.%, 0.5 wt.%, 0.2 wt.%, 0.1 wt.%, 0.05 wt.%, and 0.03 wt.%. The diluted samples as well as the stock solutions were utilized for sedimentation velocity experiments to investigate the sedimentation behavior of pure DHBC PVP-*b*-P(OEGMA). In the case of TA crosslinked PVP-*b*-P(OEGMA), the stock solution is reported in the respective paragraph.

3. Results

3.1. Synthesis of PVP-b-P(OEGMA)

The synthesis of PVP-*b*-P(OEGMA) was conducted in two steps according to the literature [59]. First, the individual homopolymers were synthesized (Figures S1 and S2); then, the blocks were coupled via CuAAc (Figure S3). Another option of block copolymer formation would be the utilization of switchable chain transfer agents, as reported recently [60]. Although PVP is a widely used polymer and is produced on a large scale, controlled polymerization is quite challenging [61]. Reversible addition–fragmentation chain transfer polymerization employing *tert*-butylhydroperoxide and sodium sulfite as redox initiators was performed at ambient temperature with an alkyne functionalized chain transfer agent to obtain alkyne end-functionalized PVP ($Đ = 1.25$; $M_n = 34,000$ g mol^{-1}, determined via SEC utilizing a PMMA calibration). On the other hand, P(OEGMA) was synthesized via atom transfer radical polymerization using 2-azidoethyl 2-bromoisobutyrate ($Đ = 1.06$; $M_n = 21,000$ g mol^{-1}, determined via SEC utilizing a PMMA calibration). Finally, the alkyne end-functionalized PVP block and the azido end-functionalized P(OEGMA) brush were conjugated via CuAAc in a mixture of water and DMSO according the literature [62]. To ensure full conversion, first, an excess of alkyne terminated PVP was used, whereas after 3 days of reaction, an azido-methyl polystyrene resin was added to the reaction mixture to capture the excess of PVP [30]. The presence of both blocks was verified via ^{1}H-NMR (Figure S3b), which shows the characteristic signal of the proton assigned to the triazole at 8.1 ppm, and thus demonstrates the success of the coupling reaction. Moreover, the successful conversion to the desired DHBC could be shown by a clear shift in the SEC molar mass distribution ($Đ = 1.42$; $M_n = 59,000$ g mol^{-1}, determined via SEC utilizing a PMMA calibration) (Figure S3c).

3.2. PVP-b-P(OEGMA) Aggregate Formation

After synthesis of the PVP-*b*-P(OEGMA), intended self-assembly in aqueous solution was investigated (Table S1). As known from the literature, the self-assembly of DHBCs in water is strongly dependent on polymer concentration [25]. To study the self-assembly behavior of PVP-*b*-P(OEGMA), DLS was conducted to identify the presence of aggregate formation on the one hand, and, on the other hand, to qualitatively compare the fraction of the unimers to the fraction of aggregates formed as a possible hint toward intended self-assembly. After dissolving PVP-*b*-P(OEGMA) in Millipore water, DLS measurements in intensity mode clearly show the formation of aggregates with apparent hydrodynamic radii in the range of 100 to 250 nm (Figure 1, Table S1), which is supported by cryo-SEM images (Figure 1 and Figure S4). DLS data in intensity mode show not only the existent self-assembly of PVP-*b*-P(OEGMA), but also the strong dependence on concentration in solution. While diluting the solution from 2.0 wt.% to 0.5 wt.%, and finally to 0.1 wt.%, the apparent fraction of the non-assembled polymers increases pronouncedly. This indicates a highly dynamic aggregate formation, depending on concentration. Furthermore, we note that the displayed intensity-weighted size distributions and obtained abundances are overestimating larger structures, which is a well-known issue of DLS analysis in solution. Thus, only a small fraction of self-assembled structures exists in solution, which can be nonetheless observed clearly in the intensity distribution. Due to the small amount of aggregates in solution, in the further course of the project, ways to improve the situation were studied. Hence, non-covalent crosslinking and stabilization of the structures was performed subsequently to investigate whether the equilibrium of aggregates and unimers can be shifted toward aggregates.

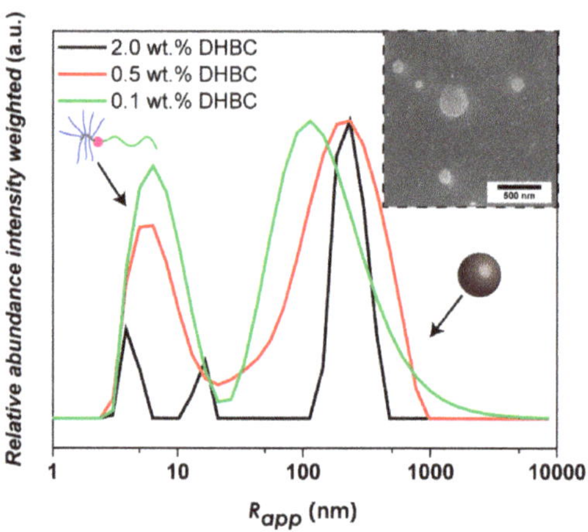

Figure 1. Intensity-weighted size distribution of PVP-*b*-P(OEGMA) in Millipore water measured via dynamic light scattering (DLS) at 25 °C and at concentrations between 0.1–2.0 wt.%. Inset: Cryogenic scanning electron microscopy (cryo-SEM) imaging of PVP-*b*-P(OEGMA) DHBC at a concentration of 2.0 wt.%, indicating existence of larger structures.

Compared to the analogous linear–linear block copolymer PVP-*b*-PEO [41], a significant increase in the apparent self-assembly efficiency, as probed by intensity-weighted DLS, can be observed (Figure 1). While in the case of a 2.5 wt.% linear–linear DHBC, DLS shows the presence of double the free polymer than of self-assembled aggregates [41]. Changing the polymer architecture to linear-brush DHBC shows a three times higher presence of aggregates than unimers at 2.0 wt.% in the DLS intensity. Therefore, a significant effect of polymer architecture on the properties can be concluded. This highlights the importance of structure–property relationships in DHBC self-assembly and aggregate formation.

3.3. Crosslinking of DHBC Aggregates via Tannic Acid

In order to shift the equilibrium between unimers and aggregates toward aggregate formation by an enhanced self-assembly stability, the aggregates were attempted to be crosslinked with TA (Figure 2a). TA is well-known for its supramolecular complexation with PVP polymers [63]. Crosslinking experiments via hydrogen bonding were conducted by diluting a 10 wt.% PVP-*b*-P(OEGMA) solution to 2.0 wt.% via a 0.5 wt.% aqueous TA solution (see SI for details). The crosslinking reaction takes place instantly, and is visually observed by a slight turbidity of the solution. Noteworthy, adding a TA solution of higher concentration (0.6 wt.%) resulted in sedimentation of the polymer–TA aggregates. This shows the strong supramolecular complexation of TA with the PVP of DHBC, while a diluted TA solution (0.3 wt.%) does not show a significant promotion of large aggregate formation. Rather, a disassembly is observed due to dilution after addition of the TA solution. Thus, a ratio of 1:7 (PVP-*b*-P(OEGMA):TA) indicates a strong non-covalent crosslinking. While diluting, the phenolic hydroxide groups undergo strong hydrogen bonding with the peptide-like functional groups of PVP located within the DHBC. Complex formation appears not only fast, but also appears stable up to the boiling point of water. This may qualify for a simple and straightforward technique to stabilize the DHBC aggregates. Non-covalent crosslinking could also be followed via Fourier transform infrared (FT-IR) spectroscopy (Figure S5). A clear broadening of the OH-stretching vibration of TA after crosslinking indicates the existence of hydrogen bonds stabilizing aggregate formation.

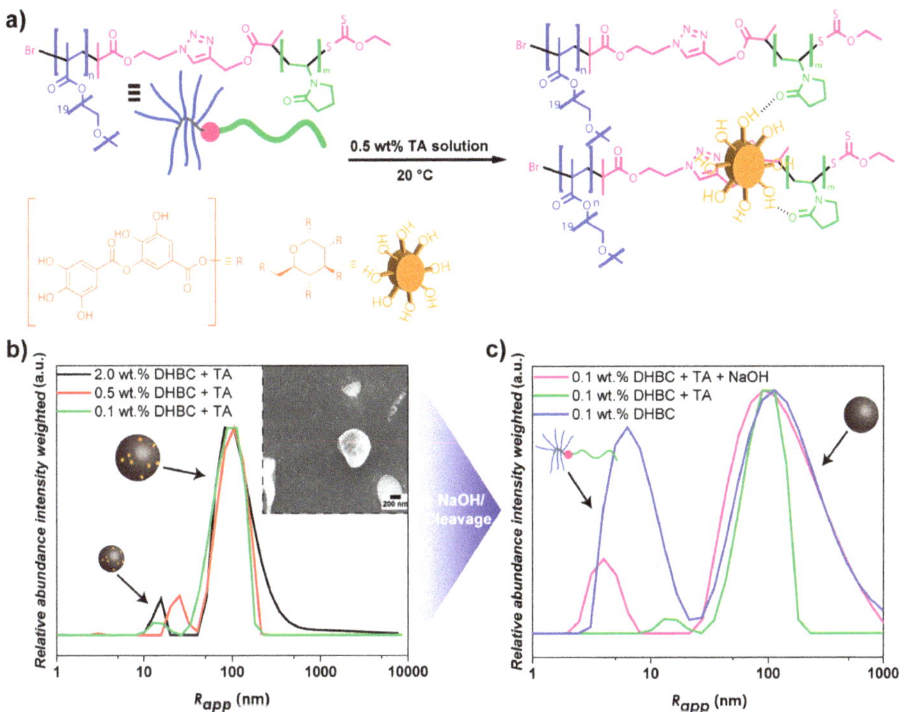

Figure 2. (**a**) Schematic overview of the non-covalent crosslinking reaction of PVP-*b*-P(OEGMA) with tannic acid (TA). (**b**) Intensity-weighted aggregate size distribution of PVP-*b*-P(OEGMA) after crosslinking with TA in Millipore water, measured via DLS at 25 °C (Inset: Cryo-SEM image of PVP-*b*-P(OEGMA) at a concentration of 2.0 wt.% after crosslinking with TA). (**c**) Intensity-weighted size distributions of PVP *b* P(OEGMA) measured via DLS at 25 °C before and after the non-covalent crosslinking procedure via TA, and the subsequent disassembly of aggregates via the addition of NaOH at 2.0 wt.%.

After the successful crosslinking of PVP-*b*-P(OEGMA) via TA, significant changes of the self-assembly could be observed via DLS in intensity-weighted size distributions (Figure 2b). Most notably, the intensity-weighted abundance of unimers decreases after the addition of TA. The self-assembly process is a highly dynamic equilibrium which leads to an ongoing self-assembly/disassembly of the unimers toward aggregates, as well as the possible exchange of unimers in aggregates by unimers in solution. The addition of TA to the DHBC solution appears to "freeze" the existent self-assembled aggregates via hydrogen bonding interactions. Moreover, the formation of smaller aggregates in size is observed, which were apparently absent before the addition of TA. The consolidated self-assembly after TA addition and stabilization of the formed aggregates can be observed in two different manners. First, prior to crosslinking, aggregate stability shows a strong concentration dependency resulting in a pronounced unimer fraction when diluted, which appears to not be the case after non-covalent crosslinking. For example, at 0.1 wt.% of polymer, almost no unimer fraction can be detected after crosslinking, but the DHBC solution shows the presence of free polymer with an apparent abundance of up to 84% according to the intensity-weighted particle size distribution. Second, the stabilizing effect of TA can also be considered from the concentration dependency of the apparent aggregate size. Before crosslinking, the particle size decreases with decreasing concentration; a similar dynamic behavior cannot be observed after crosslinking. On first sight, this may lead to a stabilized aggregate size upon dilution, indicating the non-covalent crosslinking ability of TA. Furthermore, the cryo-SEM imaging appears to underline the intensity-weighted DLS results by the presence of aggregates with apparent overall sizes around 300–400 nm (Figure 2b and Figure S6). As well, small aggregates with a hydrodynamic radius of around 15 nm can be observed, which may correspond to the smaller aggregates formed after crosslinking.

3.4. Base-Induced Disassembly of Crosslinked PVP-b-P(OEGMA) Aggregates

As TA is building hydrogen bonds with the lactam group within PVP, a change in the pH value via the addition of base leads to the suppression of such interactions, as the deprotonated TA cannot undergo hydrogen bonding with the lactam groups in PVP [64]. Nevertheless, it has been shown that the PVP–TA complex can be stable up to a pH of 10 [65]. TA crosslinked PVP-b-P(OEGMA) aggregates could be disassembled by the addition of sodium hydroxide (NaOH), as illustrated via DLS (Figure 2c and Figure S7). The self-assembly behavior of pure and TA crosslinked aggregates were compared to the self-assembly behavior of DHBC aggregates that were first crosslinked via TA and afterwards disassembled via the addition of NaOH. The samples were diluted from their initial polymer concentration (2.0 wt.%) to 0.5 wt.% and 0.1 wt.% in order to observe the re-introduced dynamics of the aggregates (Figure S7). Notably, no aggregates of pure TA were observed via DLS after deprotonation with NaOH.

While the solution of non-covalently crosslinked aggregates shows decreased amounts of unimers, and only small aggregates with a hydrodynamic radius of around 15 nm as well as large aggregates with a radius in the range of 100 nm, the addition of NaOH led to the disassembly of the small aggregates and the appearance of unimers. Therefore, the increase in pH value directly resulted in a breakdown of parts of the hydrogen bonds between TA and the PVP block. The existence of some reversibility of the crosslinking can be observed particularly after diluting the sample from the initial 2.0 wt.% to 0.5 wt.% (Figure S7) and 0.1 wt.% (Figure 2c). While the crosslinked DHBC does not show a major change in the self-assembly during dilution (Figure 2b), pure DHBC shows substantial changes as the unimer fraction is increased after each dilution step. Such dynamics are typical for non-crosslinked aggregates, as the concentration directly affects the ratio of unimers and aggregates. Aggregates that were first crosslinked via TA and afterwards treated with NaOH show a similar behavior as the unimer fraction increased with each dilution step. However, the non-covalent crosslinking is not fully reversible, as the initial fraction of unimers in pure DHBC is significantly higher than the unimer fraction after NaOH treatment (Figure 2c).

3.5. PVP-b-P(OEGMA) Aggregate Characterization via Analytical Ultracentrifugation

Another solution-based analytical method to study colloidal structures is analytical ultracentrifugation (AUC). AUC is a well-known method that enables following the sedimentation of colloidal structures in the dispersed or dissolved state. The method can provide absolute information of the colloid distribution based on typically employed concentration-sensitive detection. In recent reports, AUC failed in the identification of aggregates that were observed by the utilization of other analytical methods in studying pure DHBC and its aggregates so far [29,35]. In the present study, we utilized concentration-sensitive RI detection for the observation of sedimentation boundaries in sedimentation velocity experiments. Figure 3a clearly shows the apparent observation of a single boundary of aqueous PVP-b-P(OEGMA) solutions at 2 wt.% concentration. Notwithstanding, after non-covalent crosslinking, aggregates of DHBC were clearly identified via AUC at the same concentration (Figure 3c).

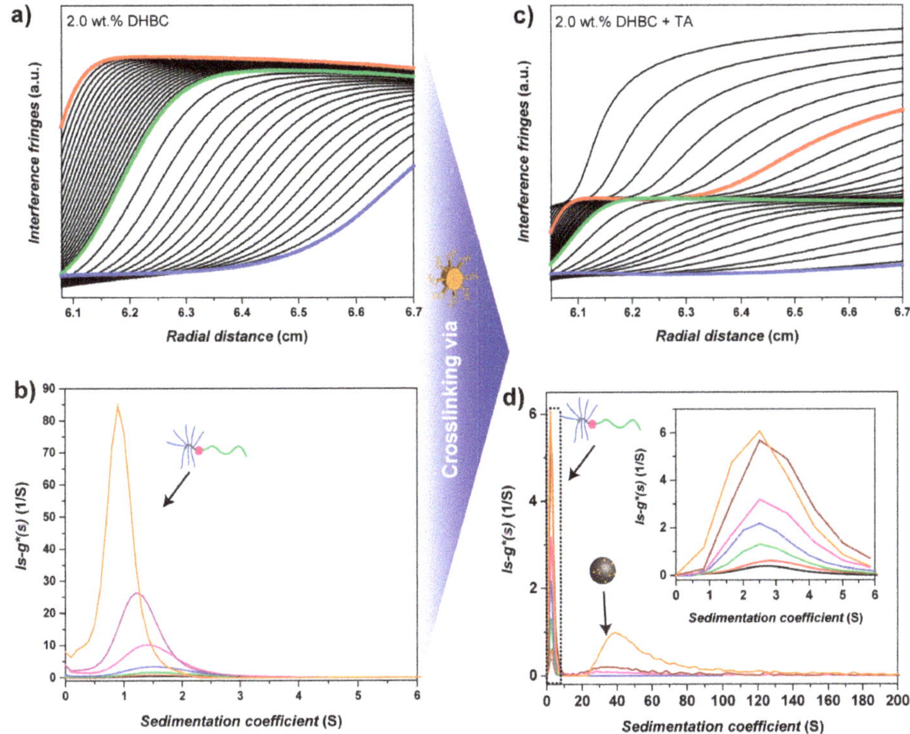

Figure 3. (**a**) Selected sedimentation velocity profiles of aqueous PVP-*b*-P(OEGMA) solution at 2.0 wt.% concentration with highlighted profiles more close to the beginning of the experiment (red), at an intermediate timescale (green), and close to the end (blue). (**b**) Sedimentation coefficient distribution, ls-g*(s), of PVP-*b*-P(OEGMA) with varying concentrations (0.03 wt.% (black), 0.05 wt.% (red), 0.1 wt.% (green), 0.2 wt.% (blue), 0.5 wt.% (magenta), 1.0 wt.% (brown), 2.0 wt.% (orange)). (**c**) Sedimentation velocity profiles of aqueous PVP-*b*-P(OEGMA) solution after the non-covalent crosslinking with TA was performed at 2.0 wt.% polymer concentration with highlighted profiles as in (**a**). (**d**) Sedimentation coefficient distribution, ls-g*(s) of PVP-*b*-P(OEGMA) after the non-covalent crosslinking reaction with various dilutions (0.03 wt.% (black), 0.05 wt.% (red), 0.1 wt.% (green), 0.2 wt.% (blue), 0.5 wt.% (magenta), 1.0 wt.% (brown), 2.0 wt.% (orange)), inset: magnification for the range of low sedimentation coefficients.

Again, the sedimentation velocity profiles of non-crosslinked PVP-*b*-P(OEGMA) has a one-step profile, which roughly corresponds to a single population of species in solution (Figure 3a). Surprisingly, we found the absence of larger structures. This could be explained by their overrepresentation in DLS-derived intensity-weighted size distributions that over-express the abundance of larger structures in solution. In most of the recent studies, intensity-weighted DLS investigations have been utilized as well, demonstrating the necessity of other in-depth studies to reconcile the dynamics in solution; most prominently, analytical ultracentrifugation failed [29,35]. The present study indicates the existence of aggregates under certain conditions. Furthermore, considering the timescale of the sedimentation velocity experiments, together with the dynamic nature and the weak hydrophilic–hydrophilic interactions of the aggregates, may make the observation of a distinct population abundance in AUC with RI detection impossible. Interestingly, decreased concentrations of the polymer (Figures 3b and 4a) led to a shift of the sedimentation coefficient distribution toward larger values, which is a strong indication of non-ideality observed at these rather high concentrations of the polymer in solution (Figures 3b and 4b) [66,67]. However, the sedimentation profiles of non-covalently crosslinked aggregates with TA show a step in the sedimentation velocity profiles. This could only be the result of two clearly distinct species moving at distinctly different speeds in the centrifugal field (Figure 3c). The apparent sedimentation coefficient distribution of suspected "unimers" and the non-covalently crosslinked aggregates is shown in Figure 3d. The first population at an apparent sedimentation coefficient of 2.5 S (signal weight average), and being largely independent of concentration appears to be the result of PVP-*b*-P(OEGMA) "unimers" (Figure 4b). The apparent second population located at ca. 60 S (signal weight average) indicates the existence of non-covalently crosslinked aggregates formed by the DHBC and TA (Figures 3d and 4b). In contrast to Figure 3b, the sedimentation coefficient distributions of species located around 3 S shows less concentration dependency (as seen by the magnification in the inset of Figure 3d and in Figure 4b). The population of species observed at higher sedimentation coefficients indicates the existence of aggregates with increased abundance at increased concentrations of polymer in solution (Figure 4a, empty circles versus half-filled circles). Interestingly, the total amount of material observed in sedimentation velocity experiments appears readily similar (Figure 4a, filled circles and filled squares), indicating the overall correctness of the mass balance.

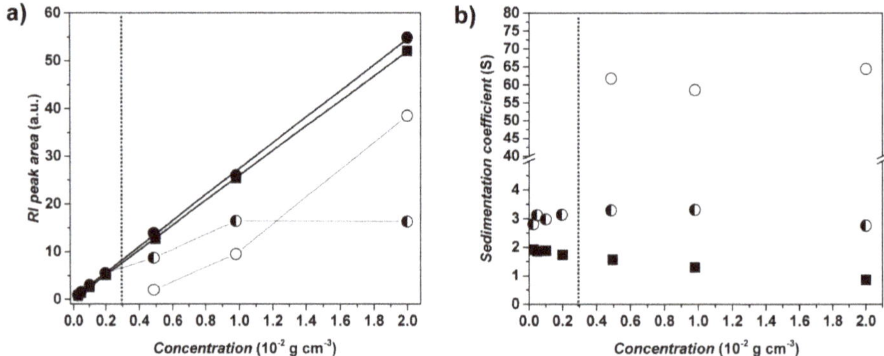

Figure 4. (**a**) Total refractive index (RI) intensities from peak areas of differential sedimentation coefficient distributions, ls-g*(s), for DHBC at varying concentrations (filled squares: without TA; filled circles: with TA) as well as RI peak areas of the population sedimenting at lower sedimentation coefficients (half-filled circles, see Figure 3d) and the abundant larger species (empty circles, see Figure 3d) appearing when non-covalent crosslinking with TA is performed. (**b**) Concentration dependence of the weight-average sedimentation coefficients (filled squares: without TA) and the two populations observed by non-covalent crosslinking with TA, sedimenting with lower sedimentation coefficients (half-filled circles, see Figure 3d) and higher sedimentation coefficients (empty circles, see Figure 3d).

Apparently, comparison of the sedimentation velocity results via AUC with DLS-derived intensity-weighted size distributions displays a modified picture of non-covalently crosslinked structures, although both methods show the occurrence of DHBC unimers and aggregates. Experiments at varying concentrations show that below 0.3 wt.%, no distinction between smaller and larger species appears possible (Figure 4) in AUC experiments. However, above 0.3 wt.%, distinction is clearly possible, with the aggregates increasing in abundance and the smaller colloid fraction decreasing in abundance. The overall mass balance of material by summing both signals appears, on first sight, only slightly affected, if at all (Figure 4a). Additionally, the apparent sedimentation velocity of the crosslinked aggregates does not show a pronounced effect on concentration (Figure 4b, empty circles), which clearly underlines their persistence, once observed in AUC. Translation of the signal (weight) average sedimentation coefficients to an apparent size is possible via the hydrodynamic equivalent sphere concept $d_h = 3\sqrt{2}\sqrt{[s]v}$ with $[s] = s\eta/(1 - v\rho_0)$ with ρ_0 being the solvent density, η being the solvent viscosity, and v being the partial specific volume of the objects in solution. v was determined via densimetry, resulting in values of $v = 0.81$ cm^{3}g^{-1} for the polymeric system without TA and

$v = 0.72$ cm^{3}g^{-1} for the system containing TA (see Figure S8). Utilization of these values leads to an apparent hydrodynamic radius of 2.5 nm for the unimers at 2 wt.% polymer without TA, 2.4 nm for the unimers, and 11 nm for the aggregates of solutions containing TA. Here, we assume solid spheres without hydration and modified friction when comparing to an ideal sphere under solution conditions at the respective concentrations. We note that the solution conditions are far away from high dilution, which is required for correct size estimations. Notwithstanding, AUC supports the findings from DLS in terms of the presence of unimers and aggregates that are similar in size when considering the volume-weighted DLS distribution (Figure S9). As indicated by DLS and cryo-SEM, the small abundance of large aggregates (in size around 100 nm) occur only in low concentrations, and thus are difficult to find in AUC measurements. This aspect is not surprising concerning the physical methodology used for deriving apparent hydrodynamic sizes in non-ideal solutions based on concentration-sensitive detection. The relatively small aggregates may also be gauged from the cryo-SEM images (e.g., Figure 2b inset and Figures S4 and S6 at high magnification).

4. Conclusions

In the present work, we could show an apparent improvement in the self-assembly process of double hydrophilic block copolymers in aqueous solution by non-covalent crosslinking with TA and by utilizing a linear-brush macromolecular double hydrophilic architecture. The improved self-assembly behavior of PVP-*b*-P(OEGMA) and its concentration dependency could be shown by DLS-derived intensity-weighted size distributions (apparent hydrodynamic radii ≈ 100–120 nm). Moreover, the DHBC assemblies could be non-covalently crosslinked via TA, which led to the additional formation of small aggregates (apparent hydrodynamic radii ≈ 15 nm). The crosslinking improved the self-assembly and apparently stabilized the aggregates upon dilution, diminishing the pronounced concentration dependency of self-assembly of the initial DHBC polymer. The presence of the aggregates could be observed via cryo-SEM before and after non-covalent crosslinking. Furthermore, particularly, the DLS results could be supported with AUC results, showing the existence of both aggregates and unimers, which was practically impossible in recent studies. Although the quantitative agreement between analytical methods is difficult to grasp momentarily, our study as well hints toward the necessity of the use of complementary analytical methods in solution to obtain a comprehensive picture of non-covalent assemblies, as the different methods provide an orthogonal clue about the physical chemistry in solution.

Supplementary Materials: The Supplementary Materials are available online at http://www.mdpi.com/2079-4991/9/5/662/s1.

Author Contributions: Conceptualization, B.V.K.J.S.; methodology, N.A.N., I.N., B.V.K.J.S.; formal analysis, N.A.N., I.N.; investigation, N.A.N., I.N.; writing—original draft preparation, B.V.K.J.S.; writing—review and editing, N.A.N., B.V.K.J.S., I.N.

Funding: N.A.N. and B.V.K.J.S. acknowledge the Max Planck Society for funding. B.V.K.J.S. thanks the Fonds der Chemischen Industrie for funding. The APC was funded by the Max Planck Society. I.N. acknowledges support of conceptual solution characterization work by the DFG-funded Collaborative Research Center PolyTarget (SFB 1278, Project Z01) and the Thüringer Ministerium für Wirtschaft, Wissenschaft und Digitale Gesellschaft (TMWWDG, ProExzellenz II, NanoPolar) for funding the Solution Characterization Group (SCG) at the Jena Center for Soft Matter (JCSM), Friedrich Schiller University Jena.

Acknowledgments: The authors acknowledge Marlies Gräwert for SEC measurements and Heike Runge for assistance with cryo-SEM. I.N. acknowledges support from U. S. Schubert.

Conflicts of Interest: The authors declare no conflict of interest. The funders had no role in the design of the study; in the collection, analyses, or interpretation of data; in the writing of the manuscript, or in the decision to publish the results.

References

1.	Zhang, L.; Eisenberg, A. Multiple Morphologies of "Crew-Cut" Aggregates of Polystyrene-b-poly(acrylic acid) Block Copolymers. *Science* **1995**, *268*, 1728–1731. [CrossRef] [PubMed]

2. Derry, M.J.; Fielding, L.A.; Armes, S.P. Polymerization-induced self-assembly of block copolymer nanoparticles via RAFT non-aqueous dispersion polymerization. *Prog. Polym. Sci.* **2016**, *52*, 1–18. [CrossRef]

3. Bang, J.; Jeong, U.; Ryu, D.Y.; Russell, T.P.; Hawker, C.J. Block Copolymer Nanolithography: Translation of Molecular Level Control to Nanoscale Patterns. *Adv. Mater.* **2009**, *21*, 4769–4792. [CrossRef]

4. Jilin, Z.; Xinhong, Y.; Ping, Y.; Juan, P.; Chunxia, L.; Weihuan, H.; Yanchun, H. Microphase Separation of Block Copolymer Thin Films. *Macromol. Rapid Commun.* **2010**, *31*, 591–608. [CrossRef]

5. Gröschel, A.H.; Walther, A.; Löbling, T.I.; Schacher, F.H.; Schmalz, H.; Müller, A.H.E. Guided hierarchical co-assembly of soft patchy nanoparticles. *Nature* **2013**, *503*, 247. [CrossRef] [PubMed]

6. Schmidt, B.V.K.J.; Wang, C.X.; Kraemer, S.; Connal, L.A.; Klinger, D. Highly functional ellipsoidal block copolymer nanoparticles: A generalized approach to nanostructured chemical ordering in phase separated colloidal particles. *Polym. Chem.* **2018**, *9*, 1638–1649. [CrossRef]

7. Ge, Z.; Xie, D.; Chen, D.; Jiang, X.; Zhang, Y.; Liu, H.; Liu, S. Stimuli-Responsive Double Hydrophilic Block Copolymer Micelles with Switchable Catalytic Activity. *Macromolecules* **2007**, *40*, 3538–3546. [CrossRef]

8. Gall, B.; Bortenschlager, M.; Nuyken, O.; Weberskirch, R. Cascade Reactions in Polymeric Nanoreactors: Mono (Rh)- and Bimetallic (Rh/Ir) Micellar Catalysis in the Hydroaminomethylation of 1-Octene. *Macromol. Chem. Phys.* **2008**, *209*, 1152–1159. [CrossRef]

9. Torchilin, V.P. PEG-based micelles as carriers of contrast agents for different imaging modalities. *Adv. Drug Deliv. Rev.* **2002**, *54*, 235–252. [CrossRef]

10. Appold, M.; Mari, C.; Lederle, C.; Elbert, J.; Schmidt, C.; Ott, I.; Stühn, B.; Gasser, G.; Gallei, M. Multi-stimuli responsive block copolymers as a smart release platform for a polypyridyl ruthenium complex. *Polym. Chem.* **2017**, *8*, 890–900. [CrossRef]

11. Zou, J.; Chen, H.; Chunder, A.; Yu, Y.; Huo, Q.; Zhai, L. Preparation of a Superhydrophobic and Conductive Nanocomposite Coating from a Carbon-Nanotube-Conjugated Block Copolymer Dispersion. *Adv. Mater.* **2008**, *20*, 3337–3341. [CrossRef]

12. Depalo, N.; Mallardi, A.; Comparelli, R.; Striccoli, M.; Agostiano, A.; Curri, M.L. Luminescent nanocrystals in phospholipid micelles for bioconjugation: An optical and structural investigation. *J. Colloid Interface Sci.* **2008**, *325*, 558–566. [CrossRef] [PubMed]

13. Schacher, F.H.; Rupar, P.A.; Manners, I. Functional Block Copolymers: Nanostructured Materials with Emerging Applications. *Angew. Chem. Int. Ed.* **2012**, *51*, 7898–7921. [CrossRef]

14. Ge, Z.; Liu, S. Functional block copolymer assemblies responsive to tumor and intracellular microenvironments for site-specific drug delivery and enhanced imaging performance. *Chem. Soc. Rev.* **2013**, *42*, 7289–7325. [PubMed]

15. Blanazs, A.; Armes, S.P.; Ryan, A.J. Self-Assembled Block Copolymer Aggregates: From Micelles to Vesicles and their Biological Applications. *Macromol. Rapid Commun.* **2009**, *30*, 267–277. [CrossRef]

16. Antonietti, M.; Förster, S. Vesicles and Liposomes: A Self-Assembly Principle Beyond Lipids. *Adv. Mater.* **2003**, *15*, 1323–1333. [CrossRef]

17. Discher, D.E.; Eisenberg, A. Polymer Vesicles. *Science* **2002**, *297*, 967–973. [CrossRef]

18. Charleux, B.; Delaittre, G.; Rieger, J.; D'Agosto, F. Polymerization-Induced Self-Assembly: From Soluble Macromolecules to Block Copolymer Nano-Objects in One Step. *Macromolecules* **2012**, *45*, 6753–6765. [CrossRef]

19. Ladmiral, V.; Semsarilar, M.; Canton, I.; Armes, S.P. Polymerization-Induced Self-Assembly of Galactose-Functionalized Biocompatible Diblock Copolymers for Intracellular Delivery. *J. Am. Chem. Soc.* **2013**, *135*, 13574–13581. [CrossRef]

20. Gaitzsch, J.; Huang, X.; Voit, B. Engineering functional polymer capsules toward smart nanoreactors. *Chem. Rev.* **2015**, *116*, 1053–1093. [CrossRef]

21. Nardin, C.; Widmer, J.; Winterhalter, M.; Meier, W. Amphiphilic block copolymer nanocontainers as bioreactors. *Eur. Phys. J. E* **2001**, *4*, 403–410. [CrossRef]

22. Rösler, A.; Vandermeulen, G.W.M.; Klok, H.-A. Advanced drug delivery devices via self-assembly of amphiphilic block copolymers. *Adv. Drug Deliv. Rev.* **2012**, *64*, 270–279. [CrossRef]

23. Blackman, L.D.; Varlas, S.; Arno, M.C.; Houston, Z.H.; Fletcher, N.L.; Thurecht, K.J.; Hasan, M.; Gibson, M.I.; O'Reilly, R.K. Confinement of Therapeutic Enzymes in Selectively Permeable Polymer Vesicles by Polymerization-Induced Self-Assembly (PISA) Reduces Antibody Binding and Proteolytic Susceptibility. *ACS Central Sci.* **2018**, *4*, 718–723. [CrossRef] [PubMed]

24. Jiang, X.; Ge, Z.; Xu, J.; Liu, H.; Liu, S. Fabrication of Multiresponsive Shell Cross-Linked Micelles Possessing pH-Controllable Core Swellability and Thermo-Tunable Corona Permeability. *Biomacromolecules* **2007**, *8*, 3184–3192. [CrossRef]

25. Schmidt, B.V.K.J. Double Hydrophilic Block Copolymer Self-Assembly in Aqueous Solution. *Macromol. Chem. Phys.* **2018**, *219*, 1700494. [CrossRef]

26. Albertsson, P.-Å. Partition of Cell Particles and Macromolecules in Polymer Two-Phase Systems. *Adv. Protein Chem.* **1970**, *24*, 309–341. [CrossRef] [PubMed]

27. Zhang, J.; Hwang, J.; Antonietti, M.; Schmidt, B.V.K.J. Water-in-Water Pickering Emulsion Stabilized by Polydopamine Particles and Cross-Linking. *Biomacromolecules* **2018**, *20*, 204–211. [CrossRef]

28. Peddireddy, K.R.; Nicolai, T.; Benyahia, L.; Capron, I. Stabilization of Water-in-Water Emulsions by Nanorods. *ACS Macro Lett.* **2016**, *5*, 283–286. [CrossRef]

29. Brosnan, S.M.; Schlaad, H.; Antonietti, M. Aqueous Self-Assembly of Purely Hydrophilic Block Copolymers into Giant Vesicles. *Angew. Chem. Int. Ed.* **2015**, *54*, 9715–9718. [CrossRef]

30. Willersinn, J.; Bogomolova, A.; Cabre, M.B.; Schmidt, B.V.K.J. Vesicles of double hydrophilic pullulan and poly(acrylamide) block copolymers: A combination of synthetic- and bio-derived blocks. *Polym. Chem.* **2017**, *8*, 1244–1254. [CrossRef]

31. Willersinn, J.; Schmidt, B.V.K.J. Self-Assembly of Double Hydrophilic Poly(2-ethyl-2-oxazoline)-b-poly(N-vinylpyrrolidone) Block Copolymers in Aqueous Solution. *Polymers* **2017**, *9*, 293. [CrossRef]

32. Oh, T.; Nagao, M.; Hoshino, Y.; Miura, Y. Self-Assembly of a Double Hydrophilic Block Glycopolymer and the Investigation of Its Mechanism. *Langmuir* **2018**, *34*, 8591–8598. [CrossRef] [PubMed]

33. Quan, J.; Shen, F.-W.; Cai, H.; Zhang, Y.-N.; Wu, H. Galactose-Functionalized Double-Hydrophilic Block Glycopolymers and Their Thermoresponsive Self-Assembly Dynamics. *Langmuir* **2018**, *34*, 10721–10731. [CrossRef]

34. Pasparakis, G.; Alexander, C. Sweet Talking Double Hydrophilic Block Copolymer Vesicles. *Angew. Chem. Int. Ed.* **2008**, *47*, 4847–4850. [CrossRef]

35. Casse, O.; Shkilnyy, A.; Linders, J.; Mayer, C.; Häussinger, D.; Völkel, A.; Thünemann, A.F.; Dimova, R.; Cölfen, H.; Meier, W.; et al. Solution Behavior of Double-Hydrophilic Block Copolymers in Dilute Aqueous Solution. *Macromolecules* **2012**, *45*, 4772–4777. [CrossRef]

36. Al Nakeeb, N.; Willersinn, J.; Schmidt, B.V.K.J. Self-Assembly Behavior and Biocompatible Cross-Linking of Double Hydrophilic Linear-Brush Block Copolymers. *Biomacromolecules* **2017**, *18*, 3695–3705. [CrossRef]

37. Cölfen, H. Double-Hydrophilic Block Copolymers: Synthesis and Application as Novel Surfactants and Crystal Growth Modifiers. *Macromol. Rapid Commun.* **2001**, *22*, 219–252. [CrossRef]

38. Hwang, J.; Heil, T.; Antonietti, M.; Schmidt, B.V.K.J. Morphogenesis of Metal–Organic Mesocrystals Mediated by Double Hydrophilic Block Copolymers. *J. Am. Chem. Soc.* **2018**, *140*, 2947–2956. [CrossRef]

39. Knop, K.; Hoogenboom, R.; Fischer, D.; Schubert, U.S. Poly (ethylene glycol) in drug delivery: pros and cons as well as potential alternatives. *Angew. Chem. Int. Ed.* **2010**, *49*, 6288–6308. [CrossRef]

40. D'souza, A.J.M.; Schowen, R.L.; Topp, E.M. Polyvinylpyrrolidone–drug conjugate: synthesis and release mechanism. *J. Controlled Release* **2004**, *94*, 91–100 [CrossRef]

41. Willersinn, J.; Drechsler, M.; Antonietti, M.; Schmidt, B.V.K.J. Organized Polymeric Submicron Particles via Self-Assembly and Cross-Linking of Double Hydrophilic Poly(ethylene oxide)-b-poly(N-vinylpyrrolidone) in Aqueous Solution. *Macromolecules* **2016**, *49*, 5331–5341. [CrossRef]

42. Al Nakeeb, N.; Kochovski, Z.; Li, T.; Zhang, Y.; Lu, Y.; Schmidt, B.V.K.J. Poly(ethylene glycol) brush-b-poly(N-vinylpyrrolidone)-based double hydrophilic block copolymer particles crosslinked via crystalline a-cyclodextrin domains. *RSC Adv.* **2019**, *9*, 4993–5001. [CrossRef]

43. Willersinn, J.; Schmidt, B.V.K.J. Pure hydrophilic block copolymer vesicles with redox- and pH cleavable crosslinks. *Polym. Chem.* **2018**, *9*, 1626–1837. [CrossRef]

44. Lehn, J.-M. Supramolecular chemistry: Where from? Where to? *Chem. Soc. Rev.* **2017**, *46*, 2378–2379. [CrossRef] [PubMed]

45. Lehn, J.-M. Supramolecular Chemistry—Scope and Perspectives Molecules, Supermolecules, and Molecular Devices (Nobel Lecture). *Angew. Chem. Int. Ed.* **1988**, *27*, 89–112. [CrossRef]

46. Wilson, A.J. Non-covalent polymer assembly using arrays of hydrogen-bonds. *Soft Matter* **2007**, *3*, 409–425. [CrossRef]

47. Schmidt, B.V.K.J.; Barner-Kowollik, C. Dynamic Macromolecular Material Design—The Versatility of Cyclodextrin-Based Host–Guest Chemistry. *Angew. Chem. Int. Ed.* **2017**, *56*, 8350–8369. [CrossRef]
48. Lohmeijer, B.G.G.; Schubert, U.S. Playing LEGO with macromolecules: Design, synthesis, and self-organization with metal complexes. *J. Polym. Sci. Part A Polym. Chem.* **2003**, *41*, 1413–1427. [CrossRef]
49. Rudolph, T.; Schacher, F.H. Selective crosslinking or addressing of individual domains within block copolymer nanostructures. *Eur. Polym. J.* **2016**, *80*, 317–331. [CrossRef]
50. Appel, E.A.; del Barrio, J.; Loh, X.J.; Scherman, O.A. Supramolecular polymeric hydrogels. *Chem. Soc. Rev.* **2012**, *41*, 6195–6214. [CrossRef]
51. Van Buren, J.P.; Robinson, W.B. Formation of complexes between protein and tannic acid. *J. Agric. Food Chem.* **1969**, *17*, 772–777. [CrossRef]
52. Chen, J.; Kozlovskaya, V.; Goins, A.; Campos-Gomez, J.; Saeed, M.; Kharlampieva, E. Biocompatible Shaped Particles from Dried Multilayer Polymer Capsules. *Biomacromolecules* **2013**, *14*, 3830–3841. [CrossRef] [PubMed]
53. Ejima, H.; Richardson, J.J.; Liang, K.; Best, J.P.; van Koeverden, M.P.; Such, G.K.; Cui, J.; Caruso, F. One-Step Assembly of Coordination Complexes for Versatile Film and Particle Engineering. *Science* **2013**, *341*, 154–157. [CrossRef]
54. Erwin, A.J.; Korolovych, V.F.; Iatridi, Z.; Tsitsilianis, C.; Ankner, J.F.; Tsukruk, V.V. Tunable Compartmentalized Morphologies of Multilayered Dual Responsive Star Block Polyampholytes. *Macromolecules* **2018**, *51*, 4800–4812. [CrossRef]
55. Kozlovskaya, V.; Zavgorodnya, O.; Chen, Y.; Ellis, K.; Tse, H.M.; Cui, W.; Thompson, J.A.; Kharlampieva, E. Ultrathin Polymeric Coatings Based on Hydrogen-Bonded Polyphenol for Protection of Pancreatic Islet Cells. *Adv. Funct. Mater.* **2012**, *22*, 3389–3398. [CrossRef]
56. Dierendonck, M.; Fierens, K.; De Rycke, R.; Lybaert, L.; Maji, S.; Zhang, Z.; Zhang, Q.; Hoogenboom, R.; Lambrecht, B.N.; Grooten, J.; et al. Nanoporous Hydrogen Bonded Polymeric Microparticles: Facile and Economic Production of Cross Presentation Promoting Vaccine Carriers. *Adv. Funct. Mater.* **2014**, *24*, 4634–4644. [CrossRef]
57. Schuck, P.; Rossmanith, P. Determination of the sedimentation coefficient distribution by least-squares boundary modeling. *Biopolymers* **2000**, *54*, 328–341. [CrossRef]
58. Quemener, D.; Davis, T.P.; Barner-Kowollik, C.; Stenzel, M.H. RAFT and click chemistry: A versatile approach to well-defined block copolymers. *Chem. Commun.* **2006**, *48*, 5051–5053. [CrossRef]
59. Li, T.; Kumru, B.; Al Nakeeb, N.; Willersinn, J.; Schmidt, B.V.K.J. Thermoadaptive Supramolecular α-Cyclodextrin Crystallization-Based Hydrogels via Double Hydrophilic Block Copolymer Templating. *Polymers* **2018**, *10*, 576. [CrossRef]
60. Pan, X.; Guo, X.; Choi, B.; Feng, A.; Wei, X.; Thang, S.H. A facile synthesis of pH stimuli biocompatible block copolymer poly(methacrylic acid)-block-poly(N-vinylpyrrolidone) utilizing switchable RAFT agents. *Polym. Chem.* **2019**. [CrossRef]
61. Pound, G.; Aguesse, F.; McLeary, J.B.; Lange, R.F.; Klumperman, B. Xanthate-mediated copolymerization of vinyl monomers for amphiphilic and double-hydrophilic block copolymers with poly (ethylene glycol). *Macromolecules* **2007**, *40*, 8861–8871. [CrossRef]
62. Bernard, J.; Save, M.; Arathoon, B.; Charleux, B. Preparation of a xanthate-terminated dextran by click chemistry: Application to the synthesis of polysaccharide-coated nanoparticles via surfactant-free ab initio emulsion polymerization of vinyl acetate. *J. Polym. Sci. Part A Polym. Chem.* **2008**, *46*, 2845–2857. [CrossRef]
63. Liu, F.; Kozlovskaya, V.; Zavgorodnya, O.; Martinez-Lopez, C.; Catledge, S.; Kharlampieva, E. Encapsulation of anticancer drug by hydrogen-bonded multilayers of tannic acid. *Soft Matter* **2014**, *10*, 9237–9247. [CrossRef] [PubMed]
64. Erel-Unal, I.; Sukhishvili, S.A. Hydrogen-Bonded Multilayers of a Neutral Polymer and a Polyphenol. *Macromolecules* **2008**, *41*, 3962–3970. [CrossRef]
65. Kozlovskaya, V.; Kharlampieva, E.; Drachuk, I.; Cheng, D.; Tsukruk, V.V. Responsive microcapsule reactors based on hydrogen-bonded tannic acid layer-by-layer assemblies. *Soft Matter* **2010**, *6*, 3596–3608. [CrossRef]
66. Grube, M.; Leiske, M.N.; Schubert, U.S.; Nischang, I. POx as an Alternative to PEG? A Hydrodynamic and Light Scattering Study. *Macromolecules* **2018**, *51*, 1905–1916. [CrossRef]

Nanomaterials **2019**, *9*, 662

67. Nischang, I.; Perevyazko, I.; Majdanski, T.; Vitz, J.; Festag, G.; Schubert, U.S. Hydrodynamic Analysis Resolves the Pharmaceutically-Relevant Absolute Molar Mass and Solution Properties of Synthetic Poly(ethylene glycol)s Created by Varying Initiation Sites. *Anal. Chem.* **2017**, *89*, 1185–1193. [CrossRef]

 *nanomaterials*

Article

Enhancing Photoluminescence Quenching in Donor–Acceptor PCE11:PPCBMB Films through the Optimization of Film Microstructure

Otto Todor-Boer [1,2,3], Ioan Petrovai [1,2], Raluca Tarcan [1,2], Adriana Vulpoi [1], Leontin David [2], Simion Astilean [1,2] and Ioan Botiz [1,*]

[1] Interdisciplinary Research Institute in Bio-Nano-Sciences, Babes-Bolyai University, Treboniu Laurian 42, 400271 Cluj-Napoca, Romania; todor.otto@gmail.com (O.T.-B.); ioan.petrovai@ubbonline.ubbcluj.ro (I.P.); raluca_tarcan@yahoo.com (R.T.); adriana.vulpoi@phys.ubbcluj.ro (A.V.); simion.astilean@phys.ubbcluj.ro (S.A.)

[2] Faculty of Physics, Babes-Bolyai University, M. Kogalniceanu Str. 1, 400084 Cluj-Napoca, Romania; leontin.david@phys.ubbcluj.ro

[3] INCDO-INOE 2000, Research Institute for Analytical Instrumentation, Donath Street 67, 400293 Cluj-Napoca, Romania

* Correspondence: ioan.botiz@ubbcluj.ro

Received: 6 November 2019; Accepted: 5 December 2019; Published: 10 December 2019

Abstract: We show that a precise control of deposition speed during the fabrication of polyfullerenes and donor polymer films by convective self-assembly leads to an optimized film microstructure comprised of interconnected crystalline polymer domains comparable to molecular dimensions intercalated with similar polyfullerene domains. Moreover, in blended films, we have found a correlation between deposition speed, the resulting microstructure, and photoluminescence quenching. The latter appeared more intense for lower deposition speeds due to a more favorable structuring at the nanoscale of the two donor and acceptor systems in the resulting blend films.

Keywords: conjugated polymers; polyfullerenes; processing by convective self-assembly; thin films and microstructure; photoluminescence quenching

---

1. Introduction

Nowadays, alternative renewable energy sources are becoming essential in our society, and organic photovoltaics (OPVs) could become in the near future a viable technological solution for continuously increasing societal energy needs. OPVs are based on semiconducting conjugated materials, and at the moment, they can convert solar energy into electricity with a power conversion efficiency of about 16.5% [1]. Researchers have shown in this last decade that in order to make efficient OPVs, one has to manipulate a series of internal fundamental phenomena that occur in the active layer such as exciton diffusion and separation [2,3], charge transport and carrier migration [4,5], or exciton and charge recombination [6] directly by controlling the microstructure of the active layer from nano- to micro- and up to the macroscale [4,6–10]. Thus, a fine control over the microstructure in the active layer is expected to tune the resulting optoelectronic properties, including the emission properties. For instance, because photoluminescence (PL) quenching is a measure of excitons separation at the donor–acceptor interfaces [11], it is important to maximize PL quenching between the donor and the acceptor materials when fabricating OPVs.

In the literature, researchers have described many efficient techniques that can be used to control more or less molecular conformations and microstructure in the active layer [12–20] and thus, to efficiently tune the emission/quenching properties [21,22]. These techniques also include convective self-assembly (CSA), which is a blade coating type technique [23] that can permit the specific control of the backbone/chain conformation by favoring more planarized polymer structures [24]. CSA is a specific methodology that is used to deposit colloidal solutions onto solid substrates. With CSA, by precisely controlling the deposition temperature and the specific tilt angle of the deposition 'blade' of the used coater, the forces acting at the triple air–substrate–solution interface can be manipulated while the solution is cast onto a carrier moving with controlled speed. As a consequence, CSA can allow us to control solvent evaporation rate and thus can be adapted to fabricate uniform, structured films using both oligomeric and macromolecular species.

Here, we employ CSA technique at various deposition speeds to produce donor–acceptor thin films of poly[(5,6-difluoro-2,1,3-benzothiadiazol-4,7-diyl)-alt-(3,3′-di(2-octyldodecyl)2,2′;5′,2′; 5′,2′-quaterthiophen-5,5′-diyl)] (PCE11; Figure 1a) and poly{[bispyrrolidino (phenyl-C61-butyric acid methyl ester)]-alt-[2,5-bis(octyloxy)benzene]} (PPCBMB; Figure 1b) that exhibit much higher PL quenching than their PCE11:PPCBMB analogues made by the spin-casting technique. As revealed by atomic force microscopy (AFM), this higher PL quenching is related to the changes in film microstructure induced upon film deposition using CSA. Our results are of importance in the OPVs field, as PCE11 and fullerenes are promising materials for organic solar cell fabrication [25–27].

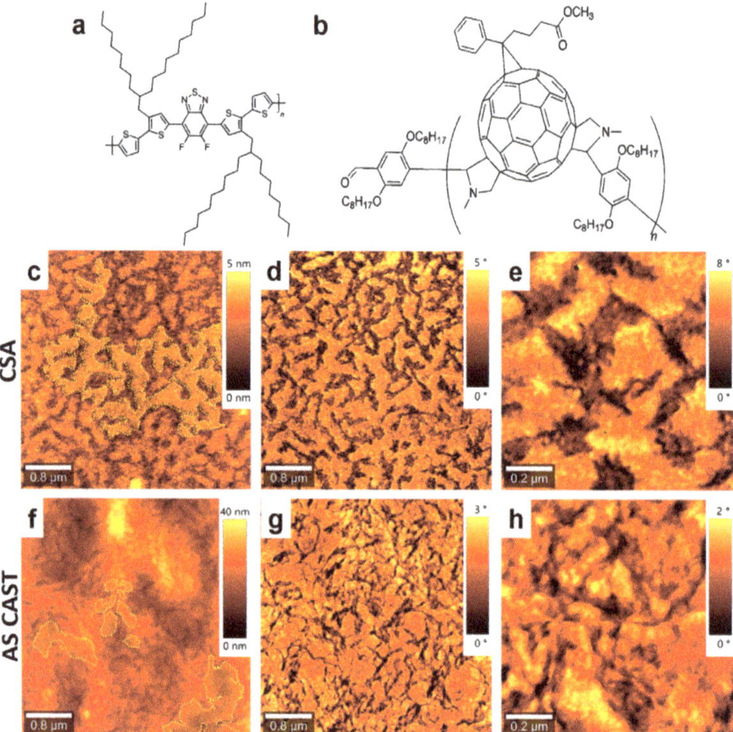

Figure 1. (**a**,**b**) Chemical structure of poly[(5,6-difluoro-2,1,3-benzothiadiazol-4,7-diyl)-alt-(3,3′-di(2-octyldodecyl)2,2′;5′,2′; 5′,2′-quaterthiophen-5,5′-diyl)] (PCE11) (**a**) and poly{[bispyrrolidino (phenyl-C61-butyric acid methyl ester)]-alt-[2,5-bis(octyloxy)benzene]} (PPCBMB) (**b**) systems. Topography (**c**) and phase (**d**) atomic force microscopy (AFM) images of a PCE11:PPCBMB thin film prepared using convective self-assembly (CSA) at a casting speed of 10 μm/s. (**e**) Zoom-in of the image shown in (**d**). Topography (**f**) and phase (**g**) AFM images of a PCE11:PPCBMB as spin cast film. (**h**) Zoom-in of the image shown in (**g**). Dotted lines are for eye guiding only and are indicating various PCE11 structures.

2. Materials and Methods

PCE11 of a weight-average molecular weight M_w = 112.707 kg/mol, number-average molecular weight M_n = 55.674 kg/mol, and polydispersity index PDI = 2.02 was purchased from Ossila Ltd. (Sheffield, UK) PPCBMB of a weight-average molecular weight M_w $\approx$ 73.8 kg/mol, number-average molecular weight M_n $\approx$ 24.6 kg/mol and poly dispersity index PDI = 3 was obtained from sterically controlled azomethine ylide cycloaddition polymerization of the phenyl-C61-butyric acid methyl ester (PCBM) as described elsewhere [28].

Thin films of PCE11 and PPCBMB were made by spin casting at 2000 rpm from 6 g/L chlorobenzene solution (with the resulting film thickness of about 80 ± 8 nm) as well as by CSA at deposition speeds of 1000 µm/s, 500 µm/s, 100 µm/s, 50 µm/s, 25 µm/s, and 10 µm/s respectively (with the resulting film thickness varying between 30 ± 5 nm and 140 ± 15 nm, respectively). A 50/50 weight percentage ratio was used to obtain solutions of PCE11:PPCBMB blend. Films of PCE11:PPCBMB were also made employing the above described procedure. For all films, regular UV–ozone cleaned microscopy cover glass was used as substrate.

The CSA coater was comprised of a motorized translation stage using a linear actuator from Zaber Technologies and moving with speeds ranging between ~4.7 µm/s and 8 mm/s. A temperature controller was placed on top of the translation stage, and the temperature of the substrate was regulated between 17 °C and 24 °C using a water-controlled system (Accel 250 LC from Thermo Scientific). A cover glass that acted as a blade was fixed in the near vicinity of the substrate at the desired angle; the polymer solution was placed on the substrate, underneath and near the edge of the blade. This configuration allowed us to precisely control both the deposition speed and the temperature of the substrate.

For the acquisition of AFM images, an Alpha 300A microscope from Witec was used in tapping mode. PL spectra were collected using an FP-6500 Spectrofluorometer from Jasco (excitation wavelength range of 220–750 nm). All PL spectra were recorded using an excitation wavelength of 640 nm.

3. Results and Discussion

Most of the work performed on OPVs has focused on film microstructures resulted by blending conjugated donor polymers with various fullerenes. Obtained microstructures most often were characterized by the existence of randomly alternating donor and acceptor domains, each greatly varying in size. Aiming to gain control over molecular packing at the nanoscale, we have decided in this work to replace fullerenes with polyfullerenes and to blend them with a crystalline PCE11 polymer system. This way, we expected to stimulate the phase separation process between polymer and polyfullerenes and thus, under specific CSA casting conditions, to control the size of resulting alternating donor–acceptor domains.

Indeed, the deposition of PCE11:PPCBMB blended films using the CSA technique led to a film microstructure that could be differentiated from that obtained in spin-cast analogue films (Figure 1). On the surface of the film prepared by CSA, brighter crystalline regions are alternating with wide darker regions (Figure 1c–e). We attribute these crystalline structures to the electron donor PCE11, which is a polymer system that is known for its high crystallinity and preferential face-on orientation of domains [25]. Note that generally, crystalline phases can be identified when using both AFM topographic (height mode) and viscoelastic (phase mode) images simultaneously [29–31]. Topography involves giving information about the surface profile/roughness while the phase is able to judge the structure of different material phases such as material softness/stiffness. As PCE11 is crystalline, i.e., stiffer material, it appears brighter in AFM phase images. Darker, most probably amorphous regions correspond to the electron acceptor PPCBMB. Moreover, these narrow crystalline domains displaying several tens of nanometers in lateral size are both randomly oriented and well-interconnected, forming complex crystalline networks (an example of such a network is delimited by dotted lines in Figure 1c) surrounded by PPCBMB regions. Instead, when analyzing the spin-cast film, we have noticed the presence of crystalline structures of various sizes ranging from very small (~0.05 ± 0.02 µm^2) to much bigger structures (~0.55 ± 0.2 µm²). No matter their size, these structures were separated from each other and surrounded by much narrower PPCBMB regions (Figure 1f–h). Therefore, they appeared rather disconnected, as indicated by the dotted lines in Figure 1f.

Furthermore, the ratio between the area of crystalline PCE11 (~57% of the total surface) and the area of amorphous PPCBMB (~43% of the total surface) regions was about 1.32 for the film produced using CSA. In comparison, for the spin-cast film, this ratio was about 4 (with 80% of the surface covered by crystalline regions and 20% of the surface covered by amorphous regions). Although AFM measurements do not exclude the existence in both films of regions with intercalated PCE11 and PPCBMB molecules, it appears that the microstructure optimized using CSA was roughly comprised of interconnected, crystalline electron donating domains displaying a width comparable to molecular dimensions that were intercalated with similar electron-accepting domains. The above described microstructure could be much altered by increasing the CSA casting speed to, for example, 1000 µm/s. In these conditions, micrometer large crystalline regions of pure PCE11 were formed, and they were separated from intermixed regions containing small structures of both PCE11 and PPCBMB systems (Figure 2).

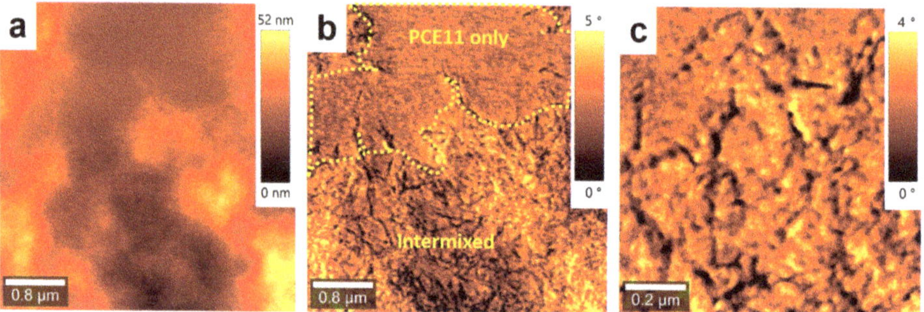

Figure 2. Topography (**a**) and phase (**b**) AFM images of a PCE11:PPCBMB thin film prepared by CSA at a deposition speed of 1000 µm/s. (**c**) Zoom-in of an intermixed region shown in (**b**).

Since PL quenching is a measure of exciton separation at the donor–acceptor interfaces, we have further compared the emission properties of the two films presented in Figure 1. We have found that there was a 65% PL quenching in the film produced using CSA compared to only a 25% PL quenching measured for the spin-cast film (Figure 3a). There are two reasons that could explain these rather low PL quenching values measured for PCE11:PPCBMB films. One reason is related to the lower unoccupied molecular orbital (LUMO) of PCE11 (−3.69 eV [25]) and of PPCBMB (−3.34 eV [28]) systems that are not necessarily favoring charge transfer (nonetheless, charge transfer conditions could still be fulfilled, because LUMO levels could be slightly modified when blending the two systems [32]). The other reason is related to the fact that long chains of polyfullerenes are more difficult to intercalate in between PCE11 polymer chains to form PCE11:PPCBMB co-crystalline domains, which could favor excitons generation/separation due to close molecular packing [32].

Moreover, PL quenching was dependent on the CSA deposition speed (and less dependent on the temperature at which the substrate was kept during casting), i.e., on the type of the resulting microstructure (Figure 3b). Thus, we attribute this improvement of PL quenching from 25% to 65% to the optimization of the film microstructure through the realization of alternating donor and acceptor domains comparable to molecular dimensions and displaying a much larger interface area that favors stronger polymer–polyfullerene quencher interactions. Similar results were obtained for several other blends when replacing PCE11 with other conjugated polymer donor systems such as poly[2,5-bis(3-alkylthiophen-2-yl)thieno(3,2-b)thiophene] (PBTTT), poly(3-(2′-ethyl)hexyl-thiophene) (P3EHT), poly (3-hexylthiophene) (P3HT), or poly(3-hexyl)selenophene (P3HS) (Figure 3c), indicating an extended applicability of the CSA technique in the processing of PPCBMB-based blends.

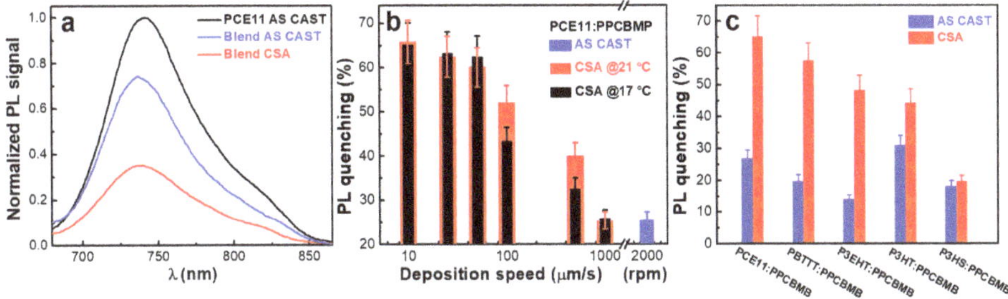

Figure 3. (**a**) Photoluminescence (PL) spectra of PCE11 and PCE11:PPCBMB films deposited by spin casting and CSA. (**b**) Amount of PL quenching in PCE11:PPCBMB films deposited by spin casting and CSA at 17 °C and 21 °C using various casting speeds. (**c**) Amount of PL quenching measured in various PPCBMB based blends that were deposited by spin casting and CSA at low casting speeds.

In order to further point out the efficiency of the CSA technique to induce structural changes in conjugated polymer/polyfullerene films, we have further investigated neat films of both PCE11 and PPCBMB. The results showed again that we can differentiate the microstructure of a PCE11 film that was deposited using CSA (Figure 4a–c) with respect to a PCE11 film that was simply spin cast (Figure 4d–f). The PCE11 film deposited using CSA at low casting speed displayed uniform granular structures that were often interconnected into rather elongated superstructures (indicated in Figure 4c by the broken lines) with a width of several tens of nanometers. Instead, the spin-cast film exhibited a microstructure that was comprised of many aggregated structures (indicated by the broken shapes in Figure 4f) randomly distributed on the surface along with much smaller granular structures.

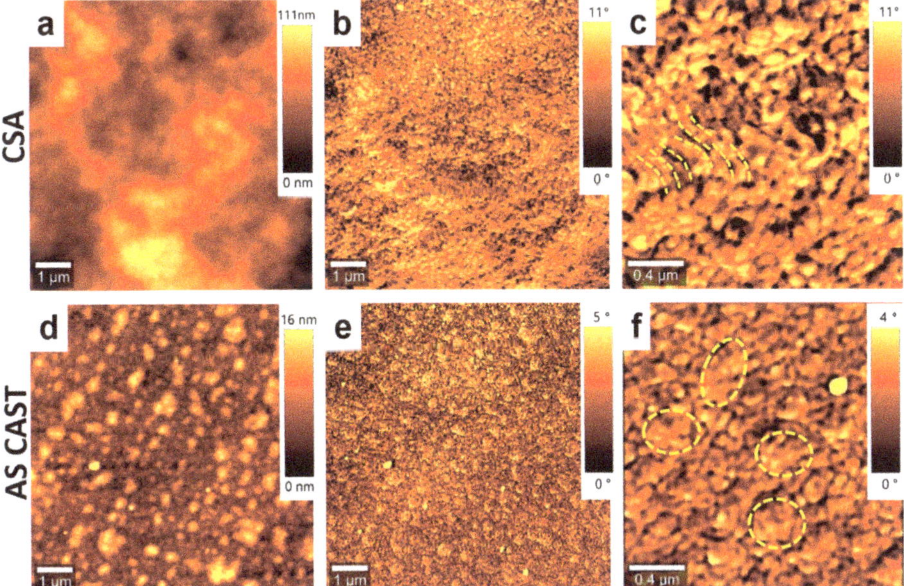

Figure 4. Topography (**a**) and phase (**b**) AFM images of a PCE11 thin film made by CSA at a low deposition speed of 10 μm/s. (**c**) Zoom-in of the image shown in (**b**). Topography (**d**) and phase (**e**) AFM images of a PCE11 as spin-cast film. (**f**) Zoom-in of the image shown in (**e**). Broken lines and shapes are for eye guiding only.

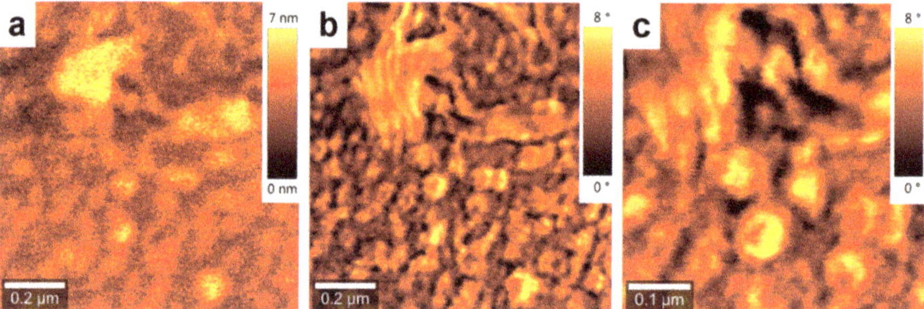

Figure 9. Topography (**a**) and phase (**b**) AFM images of a PPCBMB thin film prepared by CSA at a deposition speed of 1000 µm/s. (**c**) Zoom-in of the image shown in (**b**).

4. Conclusions

As revealed by AFM and PL spectroscopy, the quality of microstructure corresponding to thin films of PCE11 and PPCBMB depended on the casting conditions and technique. Only using CSA at low casting speed led to an optimized film microstructure that was comprised of crystalline domains of PCE11 alternated with PPCBMB amorphous regions and that exhibited strong PL quenching. These results indicate both that the microstructure of PCE11:PPCBMB films is highly sensitive to the conditions under which such films are prepared and that by carefully controlling the conditions of film preparation, we could finely tune the resulting optoelectronic properties as it is needed, for example, in OPV applications.

Author Contributions: I.B. and O.T.-B. designed the concept of paper. O.T.-B. conducted the experimental investigation and wrote the original draft. I.P., R.T., and A.V. participated in investigation. I.B., A.V., L.D., and S.A. reviewed and edited the final manuscript. I.B. and L.D. supervised O.T.-B.

Funding: This research was funded by the Romanian National Authority for Scientific Research and Innovation, CNCS – UEFISCDI, project number PN-II-RU-TE-2014-4-0013 and by the Romanian Ministry of Research and Innovation, PROINSTITUTIO project under contract nr.19PFE/17.10.2018.

Acknowledgments: We thank R. Hiorns (CNRS) for supplying PPCBMB.

Conflicts of Interest: The authors declare no conflict of interest.

References

1. Cui, Y.; Yao, H.; Zhang, J.; Zhang, T.; Wang, Y.; Hong, L.; Xian, K.; Xu, B.; Zhang, S.; Peng, J.; et al. Over 16% efficiency organic photovoltaic cells enabled by a chlorinated acceptor with increased open-circuit voltages. *Nat. Commun.* **2019**, *10*, 2515. [CrossRef] [PubMed]
2. Köhler, A.; Santos, D.A.D.; Beljonne, D.; Shuai, Z.; Brédas, J.-L.; Holmes, A.B.; Kraus, A.; Müllen, K.; Friend, R.H. Charge separation in localized and delocalized electronic states in polymeric semiconductors. *Nature* **1998**, *392*, 903–906. [CrossRef]
3. Mikhnenko, O.V.; Azimi, H.; Scharber, M.; Morana, M.; Blom, P.W.M.; Loi, M.A. Exciton diffusion length in narrow bandgap polymers. *Energy Environ. Sci.* **2012**, *5*, 6960–6965. [CrossRef]
4. Hwang, I.; Schole, G.D. Electronic energy transfer and quantum-coherence in p-conjugated polymers. *Chem. Mater.* **2011**, *23*, 610–620. [CrossRef]
5. Clarke, T.M.; Peet, J.; Nattestad, A.; Drolet, N.; Dennler, G.; Lungenschmied, C.; Leclerc, M.; Mozer, A.J. Charge carrier mobility, bimolecular recombination and trappingin polycarbazole copolymer:fullerene (PCDTBT:PCBM) bulk heterojunction solar cells. *Org. Electron.* **2012**, *13*, 2639–2646. [CrossRef]
6. Bolinger, J.C.; Traub, M.C.; Adachi, T.; Barbara, P.F. Ultralong-range polaron-induced quenching of excitons in isolated conjugated polymers. *Science* **2011**, *331*, 565–567. [CrossRef]
7. Como, E.D.; Borys, N.J.; Strohriegl, P.; Walter, M.J.; Lupton, J.M. Formation of a defect-free p-electron system in single b-phase polyfluorene chains. *J. Am. Chem. Soc.* **2011**, *133*, 3690–3692. [CrossRef]

8. Lee, P.W.; Li, W.-C.; Chen, B.-J.; Yang, C.-W.; Chang, C.-C.; Botiz, I.; Reiter, G.; Chen, Y.T.; Lin, T.L.; Tang, J.; et al. Massive Enhancement of Photoluminescence through Nanofilm Dewetting. *ACS Nano* **2013**, *7*, 6658–6666. [CrossRef]

9. Chen, W.; Nikiforov, M.P.; Darling, S.B. Morphology characterization in organic and hybrid solar cells. *Energy Environ. Sci.* **2012**, *5*, 8045–8074. [CrossRef]

10. Qin, R.; Yang, J.; Li, P.; Wu, Q.; Zhou, Y.; Luo, H.; Chang, F. Structure property relationship for carbazole and benzothiadiazole based conjugated polymers. *Sol. Energy Mater. Sol. Cells* **2016**, *145*, 412–417. [CrossRef]

11. Botiz, I.; Schaller, R.D.; Verduzco, R.; Darling, S.B. Optoelectronic Properties and Charge Transfer in Donor_Acceptor All-Conjugated Diblock Copolymers. *J. Phys. Chem. C* **2011**, *115*, 9260–9266. [CrossRef]

12. Hu, H.; Jiang, K.; Yang, G.; Liu, J.; Li, Z.; Lin, H.; Liu, Y.; Zhao, J.; Zhang, J.; Huang, F.; et al. Terthiophene-Based D–A Polymer with an Asymmetric Arrangement of Alkyl Chains That Enables Efficient Polymer Solar Cells. *J. Am. Chem. Soc.* **2015**, *137*, 14149–14157. [CrossRef] [PubMed]

13. Botiz, I.; Astilean, S.; Stingelin, N. Altering the emission properties of conjugated polymers. *Polym. Int.* **2016**, *65*, 157–163. [CrossRef]

14. Zhang, Y.; Parnell, A.J.; Pontecchiani, F.; Cooper, J.F.K.; Thompson, R.L.; Jones, R.A.L.; King, S.M.; Lidzey, D.G.; Bernardo, G. Understanding and controlling morphology evolution via DIO plasticization in PffBT4T-2OD/PC71BM devices. *Sci. Rep.* **2017**, *7*, 44269. [CrossRef] [PubMed]

15. Yang, L.; Chen, Y.; Chen, S.; Dong, T.; Deng, W.; Lv, L.; Yang, S.; Yan, H.; Huang, H. Achieving high performance non-fullerene organic solar cells through tuning the numbers of electron deficient building blocks of molecular acceptors. *J. Power Sources* **2016**, *324*, 538–546. [CrossRef]

16. Ma, W.; Yang, G.; Jiang, K.; Carpenter, J.H.; Wu, Y.; Meng, X.; McAfee, T.; Zhao, J.; Zhu, C.; Wang, C.; et al. Influence of Processing Parameters and Molecular Weight on the Morphology and Properties of High-Performance PffBT4T-2OD:PC71BM Organic Solar Cells. *Adv. Energy Mater.* **2015**, *5*, 1501400. [CrossRef]

17. Botiz, I.; Freyberg, P.; Leordean, C.; Gabudean, A.-M.; Astilean, S.; Yang, A.C.-M.; Stingelin, N. Enhancing the Photoluminescence Emission of Conjugated MEH-PPV by Light Processing. *ACS Appl. Mater. Interfaces* **2014**, *6*, 4974–4979. [CrossRef]

18. Rumer, J.W.; McCulloch, I. Organic photovoltaics: Crosslinking for optimal morphology and stability. *Mater. Today* **2015**, *18*, 425–435. [CrossRef]

19. Kolesov, V.A.; Fuentes-Hernandez, C.; Chou, W.-F.; Aizawa, N.; Larrain, F.A.; Wang, M.; Perrotta, A.; Choi, S.; Graham, S.; Bazan, G.C.; et al. Solution-based electrical doping of semiconducting polymer films over a limited depth. *Nat. Mater.* **2016**, *16*, 474. [CrossRef]

20. Cheng, P.; Yan, C.; Lau, T.-K.; Mai, J.; Lu, X.; Zhan, X. Molecular Lock: A Versatile Key to Enhance Efficiency and Stability of Organic Solar Cells. *Adv. Mater.* **2016**, *28*, 5822–5829. [CrossRef]

21. Botiz, I.; Leordean, C.; Stingelin, N. Structural Control in Polymeric Semiconductors: Application to the Manipulation of Light-emitting Properties. In *Semiconducting Polymers: Controlled Synthesis and Microstructure*; The Royal Society of Chemistry: Cambridge, UK, 2017; pp. 187–218.

22. Gagorik, A.G.; Mohin, J.W.; Kowalewski, T.; Hutchison, G.R. Monte Carlo Simulations of Charge Transport in 2D Organic Photovoltaics. *J. Phys. Chem. Lett.* **2012**, *4*, 36–42. [CrossRef] [PubMed]

23. Zhao, K.; Hu, H.; Spada, E.; Jagadamma, L.K.; Yan, B.; Abdelsamie, M.; Yang, Y.; Yu, L.; Munir, R.; Li, R.; et al. Highly efficient polymer solar cells with printed photoactive layer: Rational process transfer from spin-coating. *J. Mater. Chem. A* **2016**, *4*, 16036–16046. [CrossRef]

24. Botiz, I.; Codescu, M.-A.; Farcau, C.; Leordean, C.; Astilean, S.; Silva, C.; Stingelin, N. Convective self-assembly of π-conjugated oligomers and polymers. *J. Mater. Chem. C* **2017**, *5*, 2513–2518. [CrossRef]

25. Liu, Y.; Zhao, J.; Li, Z.; Mu, C.; Ma, W.; Hu, H.; Jiang, K.; Lin, H.; Ade, H.; Yan, H. Aggregation and morphology control enables multiple cases of high-efficiency polymer solar cells. *Nat. Commun.* **2014**, *5*, 5293. [CrossRef] [PubMed]

26. Zhang, Y.; Scarratt, N.W.; Wang, T.; Lidzey, D.G. Fabricating high performance conventional and inverted polymer solar cells by spray coating in air. *Vacuum* **2017**, *139*, 154–158. [CrossRef]

27. Zhao, F.; Li, Y.; Wang, Z.; Yang, Y.; Wang, Z.; He, G.; Zhang, J.; Jiang, L.; Wang, T.; Wei, Z.; et al. Combining Energy Transfer and Optimized Morphology for Highly Efficient Ternary Polymer Solar Cells. *Adv. Energy Mater.* **2017**, *7*, 1602552. [CrossRef]

28. Stephen, M.; Ramanitra, H.H.; Santos Silva, H.; Dowland, S.; Begue, D.; Genevicius, K.; Arlauskas, K.; Juška, G.; Morse, G.E.; Distler, A.; et al. Sterically controlled azomethine ylide cycloaddition polymerization of phenyl-C61-butyric acid methyl ester. *Chem. Commun.* **2016**, *52*, 6107–6110. [CrossRef]

29. Reiter, G.; Botiz, I.; Graveleau, L.; Grozev, N.; Albrecht, K.; Mourran, A.; Möller, M. Morphologies of Polymer Crystals in Thin Films. In *Lecture Notes in Physics: Progress in Understanding of Polymer Crystallization*; Reiter, G., Strobl, G.R., Eds.; Springer: Heidelberg, Germany, 2007; Volume 714, pp. 179–200.

30. Grozev, N.; Botiz, I.; Reiter, G. Morphological instabilities of polymer crystals. *Eur. Phys. J. E* **2008**, *27*, 63–71. [CrossRef]

31. Botiz, I.; Darling, S.B. Self-assembly of poly(3-hexylthiophene)-block-polylactide rod-coil block copolymer and subsequent incorporation of electron acceptor material. *Macromolecules* **2009**, *42*, 8211–8217. [CrossRef]

32. Jamieson, F.C.; Domingo, E.B.; McCarthy-Ward, T.; Heeney, M.; Stingelin, N.; Durrant, J.R. Fullerene crystallisation as a key driver of charge separation in polymer/fullerene bulk heterojunction solar cells. *Chem. Sci.* **2012**, *3*, 485–492. [CrossRef]

Article

Direct Measurement of Sedimentation Coefficient Distributions in Multimodal Nanoparticle Mixtures

Claudia Simone Plüisch, Rouven Stuckert and Alexander Wittemann *

Colloid Chemistry, Department of Chemistry, University of Konstanz, Universitaetsstrasse 10,
D-78464 Konstanz, Germany; simone.plueisch@uni-konstanz.de (C.S.P.); rouven.stuckert@uni-konstanz.de (R.S.)
* Correspondence: alexander.wittemann@uni-konstanz.de; Tel.: +49-(0)7531-88-5458

Abstract: Differential centrifugal sedimentation (DCS) is based on physical separation of nanoparticles in a centrifugal field prior to their analysis. It is suitable for resolving particle populations, which only slightly differ in size or density. Agglomeration presents a common problem in many natural and engineered processes. Reliable data on the agglomeration state are also crucial for hazard and risk assessment of nanomaterials and for grouping and read-across of nanoforms. Agglomeration results in polydisperse mixtures of nanoparticle clusters with multimodal distributions in size, density, and shape. These key parameters affect the sedimentation coefficient, which is the actual physical quantity measured in DCS, although the method is better known for particle sizing. The conversion into a particle size distribution is, however, based on the assumption of spherical shapes. The latter disregards the influence of the actual shape on the sedimentation rate. Sizes obtained in this way refer to equivalent diameters of spheres that sediment at the same velocity. This problem can be circumvented by focusing on the sedimentation coefficient distribution of complex nanoparticle mixtures. Knowledge of the latter is essential to implement and optimize preparative centrifugal routines, enabling precise and efficient sorting of complex nanoparticle mixtures. The determination of sedimentation coefficient distributions by DCS is demonstrated based on supracolloidal assemblies, which are often referred to as "colloidal molecules". The DCS results are compared with sedimentation coefficients obtained from hydrodynamic bead-shell modeling. Furthermore, the practical implementation of the analytical findings into preparative centrifugal separations is explored.

Keywords: nanoparticles; colloidal clusters; colloidal molecules; sedimentation; separation; classification of nanoparticles; analytical centrifugation; differential centrifugal sedimentation; disk centrifuge; density gradient centrifugation

 check for **updates**

Citation: Plüisch, C.S.; Stuckert, R.; Wittemann, A. Direct Measurement of Sedimentation Coefficient Distributions in Multimodal Nanoparticle Mixtures. *Nanomaterials* **2021**, *11*, 1027. https://doi.org/10.3390/nano11041027

Academic Editor: Federico Ferrarese Lupi

Received: 17 March 2021
Accepted: 13 April 2021
Published: 17 April 2021

Publisher's Note: MDPI stays neutral with regard to jurisdictional claims in published maps and institutional affiliations.

1. Introduction

Pursuant to the International Union of Pure and Applied Chemistry, colloidal particles are defined as objects that have "a dimension roughly between 1 nm and 1 μm" [1]. This includes the nanoscale, which is associated to the size range of approximately 1 nm to 100 nm, in line with ISO/TS 80004-2:2015 specifications [2]. Colloidal particles underlie Brownian motion. The latter prevails over sedimentation if the nanoparticles do not form agglomerates with dimensions beyond the colloidal domain. However, agglomeration and heteroaggregation of colloidal particles are major and common problems in many natural and engineered processes [3]. Surface energy will promote the agglomeration of small particles in the absence of kinetic stabilization. The latter is strongly influenced by external parameters, such as pH, salinity, or the presence of depletants [4]. Furthermore, heteroaggregation is observed when different types of colloidal particles attract each other [5]. This can be mediated by opposite surface charges, but heteroaggregation may also occur between charged and neutral particles [6]. Attachment of macromolecules to their surface may result in bridging flocculation of colloidal particles [4,7]. In cases like these, clusters of a limited number of constituent particles are formed at the onset of nanoparticle aggregation. At this early stage of aggregation, the suspension contains a multimodal mixture of assemblies with dimensions still within the colloidal regime. An accurate analytical tool to identify the onset of aggregation has thus to provide a high resolving power for mixtures of different colloidal species.

Common particle sizing techniques such as scattering techniques permit only the precise measurement of average particle sizes of monomodal colloids or bimodal mixtures that differ by at least an order of magnitude in particle size [8]. For example, static and dynamic light scattering are inherently sensitive to larger species present in a colloidal suspension. Light scattering is thus an excellent tool to monitor the onset of aggregation because the scattered light intensity is widely governed by the aggregates, while the precursor particles only contribute slightly [9]. Other techniques, such as electron microscopy, have their strength in mapping individual aggregates [10]. However, they provide poor statistics, demand specific preparation requirements, and have harsh experimental conditions. In particular, they do not permit in-situ studies.

Analytical centrifugal methods pave the way to direct studies of water-borne colloidal particles in aqueous suspension [11]. Moreover, centrifugal separation of colloidal particles provides the extraordinary resolving power required for an in-depth analysis of polydisperse nanoparticle mixtures. Two common sedimentation techniques have been established: analytical ultracentrifugation (AUC) and differential centrifugal sedimentation (DCS). AUC is a valuable tool for analyzing structural aspects of synthetic and biological nanoparticles. It is based on integral sedimentation of colloidal particles, which means that, during the analysis, the sum (the integral) of all particles smaller than a certain size is being measured [12]. Mathematical differentiation with respect to diameter will yield the (differential) particle size distribution.

Differential centrifugal sedimentation (DCS), which is also known as disk centrifuge photosedimentometry (DCP) [13] or as analytical disk centrifugation [14], enables direct measurements of differential size distributions [12]. The method combines centrifugal separation of different nanoparticle populations with continuous light extinction measurement at a fixed position. The centrifugal separation is carried out in a rotating disk (Figure 1a). The latter contains a density gradient, whose densities increase in radial direction. The gradient counteracts a higher apparent density in the sample than in layers underneath and prevents the sample from behaving as a fluid of higher density [15]. This is necessary to make the particles sediment as individual species at rates specified by their sedimentation coefficients. A detector beam is passing the density gradient at a fixed position in the vicinity of the heaviest end of the gradient. As a differential technique, only particles of a specific sedimentation coefficient reach the detector beam and are quantified by the reduction of the detector beam resulting from light scattered by particles. The sedimentation rate of a given particle population depends on its size, buoyant density, and, to a lesser extent, also on its shape [16]. It is for this reason that DCS has been widely used as a highly versatile method for particle sizing in colloid science (Figure 1c). Particle size distributions can be precisely measured down to a few nanometers [12]. This allows for sizing of synthetic and biological particles as diverse as polymer latexes [17], silica particles [13,18], gold nanospheres [18] and nanorods [19], carbon nanotubes, [20], amyloid fibrils [21], and influenza viruses [22], amongst many others.

DCS can be viewed as a complementary method to AUC, which, however, permits fast particle sizing, likewise at excellent resolution, but without the need for solving complicated equations [19]. Further refinement of the instrumentation by the manufacturer in recent years, such as optically, virtually fully transparent disks resistant to organic solvents paired with lower wavelengths of the detector beam, has made DCS a routine tool for measuring particle size distributions at high quality. In addition, substantial scientific progress was made, thereby paving the way for new directions. Armes and co-workers pioneered the analysis of various nanocomposite particles, including sterically stabilized nanoparticles [23] and raspberry-type heteroaggregates [14]. Moreover, the potential to explore colloids with surface-bound proteins was demonstrated [24,25]. Such hybrid particles are characterized by size distributions, which are superimposed by density distributions [14]. Knowledge of the particle size obtained from complementary techniques enables the determination of effective particle densities by DCS [17]. The latter are defined as the difference between the actual particle density and the density of the dispersing medium. DCS measurements thus carried out in density gradients of different composition permit simultaneous determination of particle size and density [13,22]. In the same way, combination of centrifugal sedimentation and flotation triggered by the densities of different dispersing fluids can be used for independent measurements of particle size and density [17].

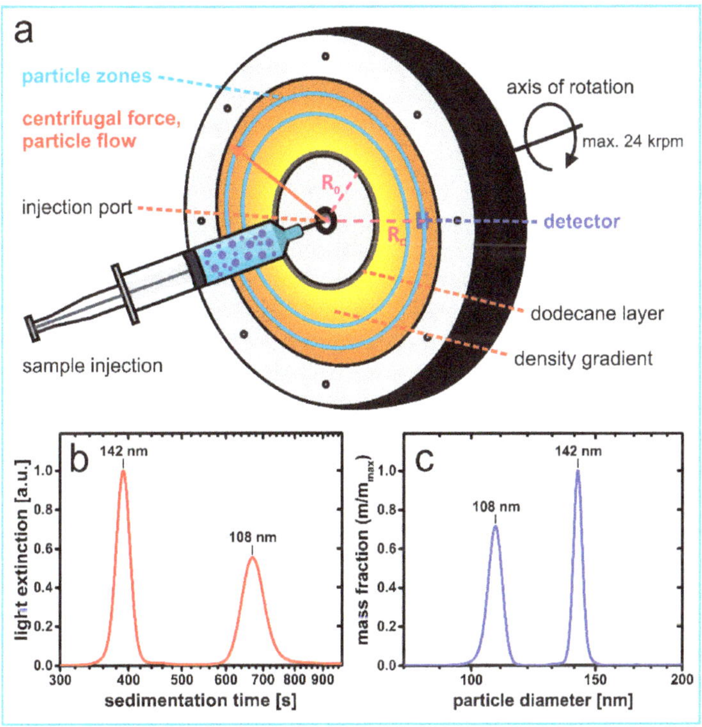

Figure 1. Differential centrifugal sedimentation (DCS) using a disk centrifuge: (**a**) A density gradient capped with dodecane is built within a rotating hollow, optically transparent disk. The sample is placed on top of the gradient via a central injection port. Different particle populations migrate as discrete zones at rates specified by their respective sedimentation coefficients. Attenuation of laser light is recorded continuously to quantify the scattering by particles arriving at a fixed detector position as a function of the sedimentation time. (**b**) Raw data of a DCS run demonstrating the high-resolving power of DCS: As an example, the analysis of a bimodal nanoparticle mixture differing by a few nanometers is shown. The latter contains spherical polystyrene latex particles with number-averaged diameters of 142 nm and 108 nm. It should be noted that the first particle population is identical to the elementary units of the clusters studied in this work, whereas the smaller particles were obtained by increasing the emulsifier concentration during particle synthesis. (**c**) For spherical particles, a particle size distribution is obtained after (i) conversion of the light extinction into particle concentrations using Mie theory, and (ii) recalculation of sedimentation times according to Equation (4).

In this article, we will demonstrate that DCS is a straightforward method to directly measure distributions of sedimentation coefficients in multimodal mixtures of nanoparticles. Colloidal clusters varying in their number of spherical constituent particles and, closely related to this, their geometries will serve as a model system. Such supraparticles are often referred to as "colloidal molecules" as their configurations resemble those of true molecules [26–28]. They have well-defined shapes, which are defined by packing criteria for uniform spheres, and can thus be regarded as prototypical for the organization of matter into nanocomposites [27]. The exemplary case study can thus be easily transferred to many other practical problems, including microscopic mechanisms underlying aggregation in aqueous nanoparticle dispersions. In this regard, DCS permits proper discrimination among species differing in their number of constituent particles and provides quick and precise access to sedimentation coefficients. The latter ones can be used to optimize centrifugal routines, enabling effective separation of nanoparticle mixtures.

2. Materials and Methods

2.1. Chemicals

Ultra-pure D(+)-sucrose ($\geq$99.9%, Proteomics Grade DNAse-, RNAse-free, VWR International GmbH, Darmstadt, Germany) and deionized water (resistivity > 18 MΩ cm) obtained from a reverse osmosis water purification system (Millipore Direct 8, Merck Chemicals GmbH, Darmstadt, Germany) were used for making aqueous density gradients. *n*-Dodecane ($\geq$99%) was purchased from Sigma-Aldrich Chemie GmbH, Steinheim, Germany and used as received. A commercial polystyrene latex standard with a particle diameter of 251 nm (HS0025-20, BS-Partikel GmbH, Mainz, Germany) was used for calibration during DCS experiments.

2.2. Colloidal Clusters

The particle clusters studied in this work were prepared by assembly of narrowly dispersed cross-linked polystyrene latex particles while confined at the surface of evaporating emulsion droplets [29]. The polystyrene particles bearing a nanometer-thin hydrophilic surface were synthesized by emulsion polymerization of styrene, divinyl benzene, and *N*-isopropylacrylamide in the presence of sodium dodecyl sulfate as the emulsifier and potassium persulfate as the initiator. A detailed description of the synthesis of the latex particles can be found in [16]. Their average diameter of 144 nm (as determined by transmission electron microscopy) and very low polydispersity index of 1.001 (given by the weight-average diameter divided by the number-average diameter) makes the spherical polymer particles perfectly suited to assembly into well-defined clusters with dimensions in the colloidal regime. The assembly into clusters was accomplished along the lines specified in [16]. The nature of the assembly hinges on trapping a limited number of particles at the surfaces of toluene droplets and packing them into clusters by strong capillary forces that occur during evaporation of the droplet phase [30]. The binding of the particles to the droplets is driven by minimization of surface energy and represents a variation of the Pickering effect with strongly swollen particles at the oil–water interface [31]. The random distribution of the polymer particles across the droplets results in a mixture of clusters differing in the number of constituents, ranging mainly from 2 to 12. In addition, the aqueous suspension of colloidal clusters also contains a fraction of single particles, resulting from droplets bearing just one particle at the surface.

2.3. Differential Centrifugal Sedimentation

Sedimentation coefficient distributions were recorded on an analytical disk centrifuge (DC24000 UHR, CPS Instruments, Inc., Prairieville, LA, USA) [12]. A density gradient was built in situ from 8.0% to 2.0% (m/m) aqueous sucrose solutions by filling the hollow disk rotating at 24,000 rpm. The gradient was covered with a thin layer of *n*-dodecane, thus minimizing evaporation of water and extending the lifetime of the gradient. The step gradient built from nine sucrose solutions (1.6 mL each) was allowed to equilibrate within 30 min, yielding a continuous gradient, which is virtually linear in volume [32].

Diluted sample suspensions (0.02% (m/m)) were sonicated to eliminate any temporary agglomerates and injected into the spinning disk (100 μL in total). The sedimentation time scale of the DCS runs is calibrated with polystyrene nanospheres as a size standard. It must be noted that this calibration is a necessary step in the software of the device before the sample measurement can be performed. However, this calibration had no influence on the actual sedimentation coefficient profiles studied in this work as the latter ones are derived from the raw data (light attenuation at $\lambda = 405$ nm vs. sedimentation time, cf. Figure 1b) and by using the single particles left in the cluster mixture as an internal reference to determine the constant k in Equation (1). By going through this procedure, the sedimentation time scale is finally defined by a component facing the same conditions as the species being investigated in the same DCS run.

2.4. Nanoparticle Separation

Preparative separation of the clusters into fractions with the same number of constituent particles was accomplished by rate-zonal density gradient centrifugation [33]. The fundamental procedure corresponds to a large extent the one reported earlier in ref. [16]. However, the focus of the current work was on aligning optimized DCS routines with preparative fractionation as tightly as possible. For this reason, several adjustments became necessary. Sucrose density gradients (36 mL), linear in volume ranging from 2% (m/m) to 8% (m/m), were prepared using a simple gradient maker, originally used for biological separations [34]. The latter is based on two chambers filled with the heaviest and lightest part of the gradient to be built. Gradual mixing of the two sucrose solutions is achieved by a rotating stir bar located in the lighter solution. Gravity results in an outward flow, which is assisted by a peristaltic pump (Masterflex L/S® Digital Miniflex® Pump, Cole-Parmer GmbH, Wertheim, Germany), enabling flowrate (0.1 mL s^{-1})-controlled gradient formation. The latter was built inside a centrifuge tube (Ultra-Clear Centrifuge Tubes, Beckman Instruments, Inc., Fullerton, CA, USA) from below using a drain tube. To facilitate the evaluation of the preparative separation, the centrifuge tubes were equipped with a scale, thus allowing precise allocation of zones of banded particle populations.

Next, 2 mL of cluster suspension (0.36% (m/m)) was carefully placed on top of the density gradient. Centrifugation was performed on an Optima XPN-90 Ultracentrifuge equipped with an SW 32 Ti swinging-bucket rotor (Beckman Coulter GmbH, Krefeld, Germany). Run conditions were 15 min at 24,000 rpm (relative centrifugal forces: $RCF_{av.} = 70{,}963$, $RCF_{max.} = 98{,}703$) and 25 °C.

The clusters separated into discrete zones were extracted from the top of the gradient using a self-built fraction recovery system. The tip of a drain tube was directly positioned into the center of the zone of banded particles to be collected. A slight negative pressure was used for gentle extraction. The cluster fractions were dialyzed exhaustively to remove sucrose prior to sample analysis (dialysis membrane: Spectra/Por® 7, MWCO 50,000 kD).

2.5. Further Methods

Transmission electron microscopy (TEM) on a Libra 120 microscope (Carl Zeiss AG, Oberkochen, Germany) at an acceleration voltage of 120 kV was used to determine the average size (144 nm) and polydispersity (1.001) of the cluster constituents based on 1000 individual particle counts. The hydrodynamic diameter of the latter is 145 nm, as demonstrated by dynamic light scattering measurements at 25 °C on an ALV/CGS-3 Compact Goniometer System (ALV-Laser Vertriebsgesellschaft m-b.H, Langen Germany). The particle density was derived by measuring the densities of a concentration series using a DMA 5000 M density meter (Anton Paar GmbH, Graz, Austria). Extrapolation of the reciprocal values of the measured densities (= specific volumes) to a solid content of 100% (m/m) yielded a particle density of 1.057 g cm^{-3} at 25 °C. It should be noted that the particle density measured by this method corresponds to the density of the bulk material [35]. However, several studies demonstrated that polystyrene latex particles have buoyant densities, which are identical within the limits of experimental errors to the density specified for the bulk material [36,37]. Fractions of colloidal clusters isolated after centrifugal separation in a density gradient were investigated by field emission scanning electron microscopy (FESEM) using a CrossBeam 1540 XB microscope (Carl Zeiss AG, Oberkochen, Germany) operating at 3 kV.

3. Results and Discussion

3.1. Analysis of Sedimentation Coefficient Distributions

Figure 1a shows schematically the basic principle of DCS. A mixture of different nanoparticle populations is physically separated into discrete zones of particles that migrate at velocities u characteristic to their size, shape, and density. This is achieved within a disk rotating at a constant angular velocity ω. Throughout the experiment, two quantities are continuously recorded. One of these is the time t it takes distinct particles to travel up a fluid density gradient from a starting position R_0 until they pass a detector position R_D. Arrival of the particles is monitored by the extinction of light at a wavelength of 405 nm. For non-absorbing materials such as polymer particles, the attenuation of a laser beam resulting from light scattered by particles arriving at specified sedimentation times is monitored.

The sedimentation coefficient s of a distinct particle population i is the quotient of the sedimentation velocity ($u = dR/dt$) and the product $\omega^2 R$, where R is the actual radial position with respect to the axis of rotation. s is inversely related to the sedimentation time t_i measured in a DCS experiment [19]:

$$s_i = \frac{u}{\omega^2 R} = \frac{\ln\left(\frac{R_D}{R_0}\right)}{\omega^2 t_i} = \frac{k}{t_i}, \tag{1}$$

Run-specific parameters such as R_0, R_D, and ω can be summarized to one instrumental measurement constant k. It is recommended to determine this constant by using a particle standard of known sedimentation coefficient. This has turned out to be beneficial if compared with making measurements by quantifying R_0, R_D, and ω individually [8].

Polymer particles with spherical shapes are particularly suitable as particle standards because their sedimentation is governed by Stokes' law [14]. A narrow size distribution will facilitate the calibration process as the local maximum of the distribution, which should correspond to the Stokes diameter of the non-agglomerated particles, can be readily determined. The current work studies mixtures of supracolloidal assemblies built from spherical constituent particles. Consequently, the latter have the same density as clusters based on them. Moreover, these particles are virtually uniform, which is reflected by their low polydispersity index of 1.001. These two conditions make the polymer nanospheres ideal reference particles to precisely determine the constant k in Equation (1). To achieve this, the sedimentation coefficient of the particles must be calculated first.

The sedimentation coefficient of a particle i is defined as the ratio of the effective particle mass m_{eff} and the friction coefficient f:

$$s_i = \frac{m_{eff}}{f} = \frac{m_p - m_f}{f}, \tag{2}$$

where the effective mass results from the difference in the actual particle mass m_p and the mass of the fluid m_f that is displaced by the particle. For spherical particles, the friction coefficient is given by Stokes' law and reads as follows:

$$f_s = 3\pi\eta d_h, \tag{3}$$

where η is the viscosity of the fluid and d_h denotes the hydrodynamic diameter of the particles. The effective mass can be expressed by the corresponding volumes of a sphere with diameter d_h and the densities of the particle $\rho_p = 1.057$ g cm^{-1} and the density of the displaced fluid ρ_f (here: 0.997 g cm^{-3}) [21]:

$$s_i = \frac{\left(\rho_p - \rho_f\right) d_h^2}{18\eta}. \tag{4}$$

The number-average diameter of the non-agglomerated particles was 142 nm, which was determined by DCS measurement against the commercial polystyrene standard latex. This value is close to the TEM diameter (144 nm) and the hydrodynamic diameter measured by DLS (145 nm).

Due to the small differences related to the low polydispersity of the particles (right peak in Figure 1c represents the size distribution), the values of any of three methods can be used as a measure for the hydrodynamic particle size. The obvious approach would be to use the DCS value of the non-agglomerated species as an internal reference to calculate the sedimentation coefficients of other species present in the mixture (see below).

Alternatively, one might prefer to calculate the masses in Equation (2) from the TEM radius characteristic for dry particles to account for the fact that the hydrodynamic effective surface layer hardly contributes to the particle mass [16]. The friction coefficient given in Equation (3) can be calculated from the hydrodynamic diameter as measured by DCS or DLS. In this study, the DLS value was used to allow for a direct comparison with sedimentation coefficients predicted earlier by hydrodynamic bead-shell modeling. In that study, model building was based on the DLS diameter of the cluster constituents [16].

An important point to note is that by using the non-agglomerated particles as an internal calibration standard, it is guaranteed that any particles present in the same DCS run, whether those to be studied or those acting as a reference, face the same conditions during their migration through the density gradient. This compensation captured in the constant k goes beyond the parameters specified in Equation (1). It also considers variations in the density and viscosity of the dispersion medium during sedimentation. This is particularly important when studying sedimentation in a density gradient. In other words, a compensation for fluid densities and viscosities that differ from the actual conditions during the experiment is achieved by using an internal reference. The sedimentation characteristics within the gradient can thus be easily transferred to the sedimentation in other dispersion media. In the following, calculation of the sedimentation coefficient of the reference particles is based on the sedimentation in pure water at 25 °C ($\rho_f = 0.997$ g cm^{-3}; $\eta = 0.891$ g m^{-1} s^{-1}). The latter is done not only for the sake of simplicity, but also to allow for direct comparison to predicted values from hydrodynamic modeling that have been reported previously [16]. Consequently, a sedimentation coefficient of $s_{N=1} = 770$ Sv is calculated from Equation (4). The sedimentation time of the nanospheres assigned from the DCS analysis of cluster mixtures is $t_{N=1} = 401$ s, which yields a k-value of 3.0877×10^{8} according to Equation (1). This value is then used to calculate the sedimentation coefficients of any non-spherical species present in the nanoparticle mixture from their respective sedimentation times.

3.2. Particle Clusters as Model Systems for Nanoparticle Mixtures

Thanks to the seminal work by Pine and co-workers [30], it is known that evaporating emulsion droplets are suitable physical templates to assist the assembly of colloidal particles into an ensemble of stable clusters varying in their number of constituents. A mixture of various species is obtained, resulting from a random distribution of the elementary particles on the droplets [31]. The gradual increase in cluster mass follows directly from the number of constituent particles N. A linear correlation does not apply to the increase in surface area and, related to this, friction at the cluster surface. The growth in surface area is largest when going from single particles to dimers, followed by increasingly smaller growth rates with rising N. The latter is a direct consequence of the packing into dense cluster configurations. It is obvious that the sedimentation coefficients of the clusters gradually increase with rising N in accordance with Equation (2). Moreover, based on the above, the increments for the rising sedimentation coefficients are becoming increasingly smaller with rising N. This fact underscores the suitability of particle clusters as model systems to explore the limits of sorting nanoparticle mixtures by centrifugal sedimentation.

The following studies are based on a mixture of particle clusters built from 144 nm-sized polymer nanospheres (Inset of Figure 2). Most of the clusters consist of up to 12 constituent particles and thus have spatial dimensions within the colloidal domain. Hence, Brownian motion will prevail over sedimentation unless the assemblies are exposed to a centrifugal field [38]. The morphology of the colloidal clusters in this work and of related supraparticles was discussed in earlier works [16,28,30,31,39,40]. It should be noted that clusters of more than four constituents may occur in different configurations. For example, clusters of five and six particles have two different configurations. Figure 3 shows common configurations that are observed experimentally and predicted by computer simulations. The cluster mixture contains both isotropic and anisotropic species. Single particles, tetrahedral, octahedral, and icosahedral clusters have isotropic geometries, whereas all other species exhibit anisotropic shapes. The further studies will thus deliberately dispense on any evaluations based on assuming a spherical particle geometry. The basic methodology is thus broadly applicable to particles of arbitrary shapes that do not have to obey Stokes' law.

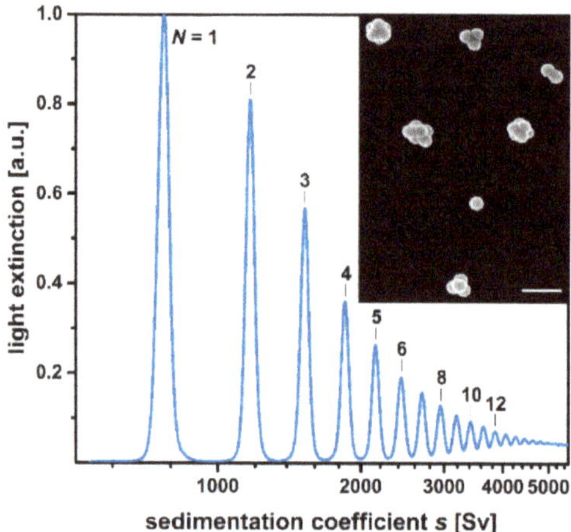

Figure 2. Sedimentation coefficient distribution of a mixture of colloidal clusters as measured by DCS. Cluster species of more than 12 constituent particles are resolved separately as narrow bands. DCS analysis thus enables a clear classification of nanoparticles according to their sedimentation characteristics. The inset shows an FESEM micrograph of the cluster mixture studied by DCS. The scale bar represents 400 nm.

3.3. DCS Analysis of Sedimentation Coefficient Distributions

DCS is now applied as an analytical tool to explore the distribution of sedimentation coefficients of the mixture of colloidal clusters. The separation of the particles within the disk centrifuge follows the basic principles of rate-zonal density gradient centrifugation [15]. The proper design of the density gradient is key for accurate results. The following criteria must be considered when preparing the density gradient:

1. The density of the particles must exceed the highest density within the gradient.
2. The density of the sample suspension (particles + dispersion medium; here: 0.997 g cm^{-3}) has to be lower than the lowest density within the gradient.

N	s_{DCS}	s_{PC}	s_{BS}	[Sv]	configuration
1	770	829	770		single particle
2	1170	1261	1166		doublet
3	1521	1607	1536		triplet
4	1849	1926	1911		tetrahedron
5	2144	2192	2164		square pyramid
5	2144	2192	2155		triangular dipyramid
6	2431	2459	2601		octahedron
6	2431	2459	2571		pentagonal dipyramid minus one
7	2685	2701			pentagonal dipyramid
8	2941	2918		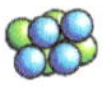	snub disphenoid
9	3183	3138			triaugmented triangular prism
10	3393	3332			gyroelongate square dipyramid
11	3633	3529		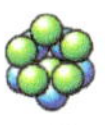	icosahedron minus one
12	3860				icosahedron

Figure 3. Sedimentation coefficients of the cluster species varying in the number of constituent spheres N and, consequently, their configurations. Analytical values s_{DCS} obtained from DCS are compared with sedimentation coefficients s_{PC} estimated from preparative separations by rate-zonal density gradient centrifugation. In addition, sedimentation coefficients s_{BS} predicted from hydrodynamic bead-shell modeling, as taken from [16], are given for clusters with $N \leq 6$. Model building considers the exact geometries of the colloidal clusters. Notwithstanding the shortcomings in deriving sedimentation coefficients from preparative centrifugations, the quantities thus obtained differ by less than 8% from the analytical values and the theoretical predictions.

Compliance with the two criteria makes the particles settle individually at rates specified by their sedimentation coefficients. In the present case, clusters of a single set of constituent particles were explored at 25 °C. Hence, the various species within the cluster mixture have the same density, which is 1.057 g cm^{-3}. Their separation during DCS analysis was performed in an aqueous sucrose gradient ranging from 2% (m/m) to 8% (m/m). The minimum density within the gradient equals 1.0052 g cm^{-3} and is thus, in accordance with criterion 2, higher than the density of the highly diluted sample suspension (0.02% (m/m), $\rho = 0.997$ g cm^{-3}). This prevents streaming, i.e., a downstream of particle-laden fluid following the centrifugal field [15]. In addition, the maximum density within the gradient (1.0285 g cm^{-3}) was kept lower than the actual particle density. This ensures that the particles can travel though the complete gradient fluid and will thus reach the detector position at a characteristic sedimentation time. The gradient design suitable for polystyrene latex particles can be easily adapted for other types of nanoparticles, simply by following the two criteria. The densities of aqueous sucrose solutions as functions of weight fraction and temperature can be found in the literature [41]. Figure S5 displays densities in the concentration range up to 10% (m/m) sucrose.

The particle clusters do not absorb visible light, but light is scattered by the clusters, as for any other colloidal objects. The colloidal clusters can be thus detected by attenuation of a laser beam. The latter had a wavelength of 405 nm, which provides high sensitivity because violet light is subject to the strongest scattering within the visible region. For this reason, even marginal quantities of particles or particle agglomerates can still be detected.

During the DCS run, the light attenuation of the detector beam is continuously recorded as the function of the sedimentation time. Routines in the device software allow for conversion of the raw data into a particle size distribution. These routines are based on the Mie theory to derive the particle concentration from the measured light extinction [42]. Moreover, the device software assigns a distinct particle diameter to any given sedimentation time. However, this correlation is only correct if the drag force acting on the particles during sedimentation follows Stokes' law. The latter applies only to spherical particles and is thus not appropriate for accurately exploring particles with complex shapes. The deviation between the friction coefficients of an anisotropic particle and a spherical particle of the same mass can be at least partially compensated by the introduction of a non-sphericity factor [12]. However, this can only work if all the particles exhibit the same aspect ratio, which is rarely the case for mixtures of anisotropic particles. Nanoparticle aggregates such as the colloidal clusters in this study have various geometries (Figure 3), so that the assignment of a single non-sphericity factor that works for all species is not possible.

It is exactly for this reason that the present evaluation explicitly avoids any assumptions of a spherical shape and is thus applicable to particles of arbitrary shapes. The only exception is made when calculating the sedimentation coefficient of the spherical particles used as an internal reference. In this specific case, making use of Equation (3) is justified. Basically, it is also possible to replace the spherical particles by any other type of particles given that their sedimentation coefficient is known. Precise knowledge of the sedimentation coefficient of at least one set of reference particles is required to determine the instrumental measurement constant k in Equation (1). These reference particles can be either measured independently of the sample particles, or, as in the present case, constitute a distinct particle population within the nanoparticle mixture. The latter procedure is advantageous as different particle populations are compared, which are subject to the same experimental conditions. In case of nanoparticle aggregates, it makes sense to use the elementary particles as an internal reference. The k-value calculated from the sedimentation time of the reference particles and their sedimentation coefficient can thus be transferred to any other particle population within the same DCS run. Sedimentation coefficients of the different particle populations are calculated from the respective sedimentation times according to Equation (1).

Together with the profile of light extinction measured during DCS, a distribution of the sedimentation coefficients of the nanoparticle mixture is obtained. This distribution is weighted by the attenuation of light caused by the particle arriving at the detector position and thus differs from the weight distribution $c(s)$ obtained from AUC [43]. In principle, conversion of the two contributions is possible if the scattering cross-sections of the various particle populations are known. According to the authors' opinion, this issue should be treated in a purist way to allow for a broad application of the methodology.

Figure 2 shows the light extinction-weighted sedimentation coefficient distribution of a mixture of colloidal clusters. As described above, the distribution is solely based on the raw data of the DCS run and knowledge of the sedimentation coefficient of the single particles used as an internal reference. The high resolution of the DCS run is immediately obvious. Single particles and cluster species of 2 to 12 constituents are resolved as discrete bands, which allows for a facile assignment of sedimentation coefficients according to the peak maxima (Figure 3). Remarkably, species that deviate by less than 1% in their sedimentation coefficients can still be resolved. A more detailed discussion of the resolution of DCS is found in ref. [12].

Zone widths in a DCS experiment are governed by several factors. Because DCS detects particles arrive at a fixed position over time, the actual zone width of a given particle population is captured as a time interval within which all particles of this population arrive at the detector position. A broad zone is thus reflected by a larger time interval. This becomes apparent when the light extinction is plotted against the sedimentation time (Figure S1). The peak-width at half-height is largest for the non-agglomerated particles (25.2 s) and decrease systematically for the clusters with rising numbers of constituent particles N (13.0 s for dimers; 9.2 s for trimers; 7.3 s for tetramers; 6.0 s for pentamers; 5.4 s for hexamers). The polydispersity of the cluster constituents (1.001) and, consequently, of the clusters is rather low and virtually negligible compared to the broadening due to Brownian motion. Band broadening due to diffusion is an important aspect. Translational diffusion coefficients decrease with the number of constituent spheres [16,38] and, consequently, facilitate band narrowing with rising numbers of constituent particles N (Figure S1). The density gradient influences the band width as well. A steep gradient profile results in lower sedimentation velocities in the leading edge of the zone, whereas the velocities are higher at the trailing edge. The density gradient thus keeps the zone of a distinct particle population together. The density gradient equally affects all particle populations because the detector position is fixed in DCS measurement. This constitutes a difference from preparative centrifugal separations, where different particle populations are harvested from different sections of the density gradient (Figure 4).

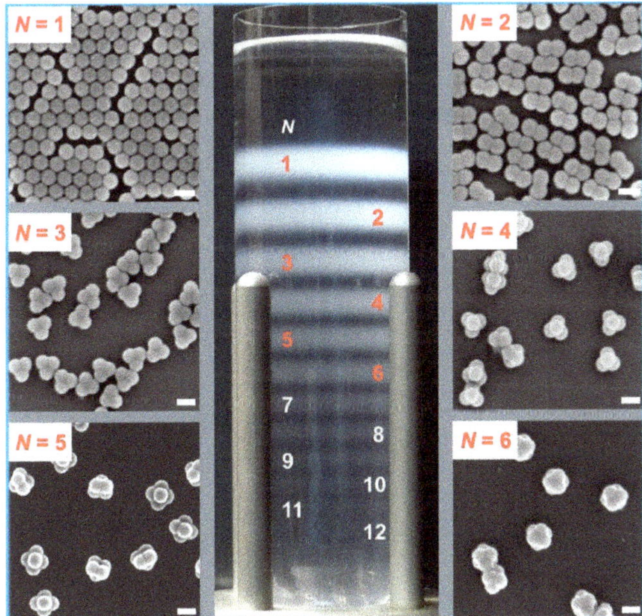

Figure 4. Centrifugal separation of colloidal clusters according to their sedimentation coefficients in a sucrose density gradient ranging from 2% (m/m) to 8% (m/m). The fractionation was carried out in a swing-out rotor at 24,000 rpm. Cluster populations of up to 12 constituent particles were isolated as individual zones that could be harvested by using a self-built fraction recovery unit. FESEM micrographs of fractions of particle monomers (N = 1), dimers (N = 2), trimers (N = 3), tetramers (N = 4), pentamers (N = 5), and hexamers (N = 6) are grouped around the centrifuge tube hosting the gradient and the particle zones. The scale bars represent 200 nm.

According to Equation [2], there is an inverse relationship between sedimentation time and sedimentation coefficient. This must be considered when comparing actual zone widths during the DCS run with the widths of the individual peaks in the sedimentation coefficient distribution (Figure 2). It follows that the non-agglomerated particles which form the broadest zone (peak-width at half-height: 25.2 s) during the DCS run are presented by the narrowest peak in the sedimentation coefficient distribution (peak-width at half-height: 48 Sv). In comparison, the tetramers form narrower zones (peak-width at half-height: 7.3 s) but are represented by a broader peak in the sedimentation coefficient profile (peak-width at half-height: 81 Sv).

Accurate assessment of the sedimentation coefficients obtained from the DCS strategy can be based on a comparison with values obtained from theoretical modeling. García de la Torre and co-workers have established model building and calculation routines, which can be applied to rigid particles of arbitrary shapes. Recently, we reported on sedimentation coefficients of colloidal clusters calculated along these lines [16]. The clusters studied are identical to those in the present studies but limited to a maximum of six constituent particles. Nonetheless, a direct comparison can be made (Figure 3). The sedimentation coefficients obtained by DCS are in excellent correlation to the values predicted for clusters with the same spatial dimensions and geometries. Deviations are negligible for single particles and even for dimers, which are the geometries with the highest aspect ratio. Although the deviations increase with the number of constituents, they are less than 6% for six-particle clusters with octahedral symmetry. This underscores the suitability of DCS to directly measure sedimentation coefficient distributions of complex nanoparticle mixtures.

3.4. Application for Preparative Nanoparticle Separations

The separation quality accomplished during the DCS analysis should now serve as a basis to optimize preparative centrifugal separations. To this end, separations according to the sedimentation coefficients were carried out in a swinging-bucket rotor. The buckets host clear centrifuge tubes with a capacity of 38.5 mL. Compared to a fixed-angle centrifuge rotor, swing-out rotors significantly reduce collisions of the particles with the wall of the tube and help to reduce wall effects on sedimentation (see below). More importantly, the fluid layers and the particle zones follow the centrifugal field throughout the whole run, including acceleration and deceleration intervals, which is the prerequisite for excellent selectivity during fractionation. The orientation of the layers with respect to the axis of rotation is a feature which DCS analysis and separations carried out in a swing-out rotor have in common. In this context, the spinning hollow disk in DCS can be considered as an extension of the centrifuge tube in a swing-out rotor by 360 °C (Figure 1).

Nevertheless, there are two important differences that should be kept in mind. The first one is related to the geometry of the disk, which has some similarities to a zonal rotor used for preparative separations at larger scales [32]. A density gradient linear in volume will be also linear along the sedimentation path if placed in a centrifuge tube. However, if prepared within the rotating hollow disk, a concave profile of the density gradient is obtained across the sedimentation path. The second difference to be considered relates to the substantial difference between an analytical technique and a preparative method. DCS probes the different particle populations after having reached a fixed position. This means that all particles have migrated along the same path, but they have reached their destination at different times according to their sedimentation coefficient. In preparative separations, the principle of operation is reversed. The centrifugal run and thus the sedimentation are terminated for all particle populations at the same time. At precisely this time, they have traveled different distances, which are determined by their sedimentation coefficients. Hence, the two methods offer the opportunity to complement each other if aligned at the same experimental problem.

Parallel to the above-mentioned DCS analysis, sucrose density gradients ranging also from 2% (m/m) to 8% (m/m) were prepared in centrifuge tubes. In line with the DCS analysis, the centrifugation was performed at 24,000 rpm. Centrifugation times were chosen based on the sedimentation coefficients measured by DCS. In doing so, the time required to maximize the sedimentation path of a 12-particle cluster was estimated. This ensured maximum separation of the cluster populations.

Figure 4 shows the separation achieved in the centrifugal run. A total of twelve discrete zones, each of them corresponding to a particle population settling at the same rate, are observed. Classification of the zones to specific clusters follows the number of constituent particles N. This is in full accord with the sequence of sedimentation coefficients obtained from DCS (Figure 3). Verification of zone allocation was achieved by FESEM after extraction of the cluster fractions (Figure 4).

The separation of the cluster mixture into discrete zones within the centrifuge tube paved the way for a quantitative evaluation. There are a number of reports in the literature on the calculation of sedimentation coefficients from separations of biomacromolecules performed in a centrifuge tube [15,44,45]. According to Equation (1), it is possible to calculate sedimentation coefficients from the radial locations of the various zones once the centrifugal run is terminated. To this end, a centrifugal separation of the cluster mixture was carried out in a graduated test tube. The R_D values in Equation (1) are taken from the center of mass of the zones. The initial position R_0 is assumed as the interface between sample zone and density gradient (Figure S3). The effective time of centrifugation was determined along the lines given by Schumaker [15]. To this end, the course of the rpm values during the entire centrifugal run was recorded. Periods of acceleration and deceleration are considered by plotting the angular velocity-squared ω^2 with the course of the centrifugation time (Figure S2). Integration of ω^2 over the entire time span and division by the maximum value of ω_{max}^2 gave an effective centrifugation time of 470.4 s at maximum angular velocity ω_{max} (here: 2513.4 s^{-1}).

The calculation according to Equation (1) will yield the actual sedimentation coefficients of the different nanoparticle populations within the density gradient. The latter can be considered as dispersion medium with a gradual change not only in density but also in viscosity. In DCS, the impact of the medium is considered by the measurement constant k. Hence, sedimentation coefficients determined by DCS reflect experimental conditions identical to those assumed for the particles used as a calibration standard. The data gathered in Figure 3 refer to the sedimentation in pure water at 25 °C. To allow for a comparison with the DCS values, it is necessary to correct the values determined from preparative separations by the mean viscosities and densities, which the particles experience during their sedimentation. This correction can be made by using the following expression [44]:

$$s_{PC} = s_G \cdot \frac{\rho_p - \rho_W}{\rho_p - \rho_G} \cdot \frac{\eta_G}{\eta_W}, \tag{5}$$

where s_{PC} and s_G denote the sedimentation coefficients of a given particle population at 25 °C in pure water and in the gradient. $\rho_W = 0.997$ g cm^{-3} and $\eta_w = 0.891$ g m s^{-1} are the density and viscosity of water at 25 °C.

The mean density and viscosity of the gradient, ρ_G and η_G, were determined as follows: first, the sucrose concentrations at the zone centers were calculated based on the linear profile of the gradient (Figure S3). Densities and viscosities of aqueous sucrose solutions given in ref. [41] were plotted against the weight proportion of sucrose and subjected to a polynomial fitting (Figures S4 and S5). The fitting functions were integrated from the minimal concentration of sucrose in the gradient (2% (m/m)) to the concentration of sucrose at the zone center (Figure S3). Division by the difference in the two sucrose concentrations yields the mean density and viscosity during the sedimentation of a given particle population. In the concentration range considered, the densities and viscosities increase virtually on a linear scale (Figures S4 and S5). Hence, the derivation of the quantities ρ_G and η_G could be also simplified by averaging the respective quantities at the beginning and at the end of the sedimentation path.

Figure 5 shows a comparison of the sedimentation coefficients s_{PC} calculated from the radial positions of the zone centers within the centrifuge tube with the quantities determined by DCS. Notably, the sedimentation coefficients obtained by the two strategies deviate by less than 8%. This is remarkable, inasmuch as there are a number of uncertainties in the calculation of accurate sedimentation coefficients from the results of rate-zonal separations carried out in centrifuge tubes [46]. The following issues must be considered:

1. uncertainty in the particle density;
2. uncertainties in the profile of the density gradient;
3. uncertainties in the positions of the zones;
4. temperature variations inside the rotor chamber;
5. wall effects.

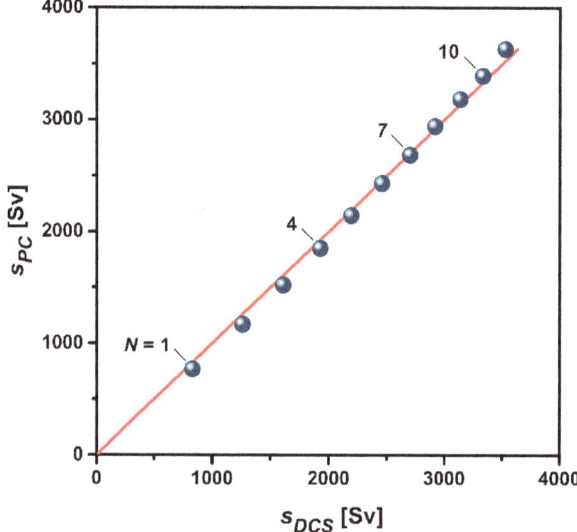

Figure 5. Cross-comparison of sedimentation coefficients shows excellent agreement between the quantities determined by DCS and those estimated from preparative centrifugal separation. The bisecting red line is added for ease of classification and to indicate deviations from full convergence.

This study has shown that deviations in comparison to other methods can be kept low. It is briefly outlined how this was achieved despite the uncertainties listed above.

This study is centered on the determination of the sedimentation of solid nanoparticles. In contrast to most biological particles or synthetic micro- or nanogels, solid particles do not vary their density while migrating through a density gradient. Uncertainty in particle density is thus limited to experimental errors in the determination of particle densities. The latter can be precisely measured with an uncertainty of less than 0.005 g cm^{-3}.

Considerable uncertainties in the density gradient profile shown in Figure S3 are to be expected in the top layer of the gradient underneath the sample zone. Diffusion of gradient material (here: sucrose) into the sample zone may cause strong deviations from the expected gradient profile [32,46]. However, it turned out that this phenomenon had little effect on the calculated values of the sedimentation coefficients. This was achieved by setting an effective sedimentation time, allowing the zones of migrating particles to reach positions far from the first layers of the gradient (Figure S3).

Uncertainties in radial positions relate to both the starting zone and the final positions of the zones of banded particles. Precise allocation of the beginning of the sedimentation path is challenging inasmuch as the profile of density gradient is not well known at the interface between the sample zone and the top layer of the gradient. It may thus be justified to define the starting position either by the interface itself or as the center of mass within the sample zone. In the present case, the two positions differed by 2.1 mm. This value can be reduced by smaller sample volumes at the expense of the number of particles that can be separated in a single run. In this work, the nanoparticles were dispersed in pure water. For this reason, streaming is to be expected within the sample zone. This is why the diverse nanoparticle populations rapidly accumulate near the interface of sample and gradient during acceleration of the centrifuge. It is exactly for this reason that the radial position of the beginning of the gradient ($R = 70.14$ mm) was chosen as the beginning of the sedimentation path for all nanoparticle populations. This definition was corroborated by the agreement with the values from DCS and hydrodynamic modeling.

The sedimentation coefficients were calculated from the radial positions of the centers of mass of the zones, which are affected by the centrifugal field. As a result, the particle distributions deviate from Gaussian profiles. In the present case, uncertainty of the zone position had little effect because of the low bandwidths resulting from the narrow size distributions of the particle populations. Nonetheless, further refinement could be achieved by taking into account the interplay of sedimentation and diffusion along the lines given by Schumaker [15].

Variation in the temperature inside the rotor chamber may have a marked effect on the sedimentation of the particles. An uncertainty of 1 °C will cause an error of over 2.5% in the determination of sedimentation coefficients [46]. Variations in temperature affect both the density and viscosity of the gradient and can become a major problem when calculating sedimentation coefficients from centrifugal separations. In the DCS measurements shown above, this problem was circumvented by using an internal calibration standard of known sedimentation coefficient. It is also possible to use the same strategy in preparative centrifugal separations. Alternatively, values obtained from preparative separations can be corrected by DCS analysis of at least one component. In the present case, this was not done to identify deviations among the methods.

Collisions of nanoparticles with the walls of cylindrically shaped centrifuge tubes may occur. Particles that hit the wall may stick to it or accumulate near the wall and then settle down as an ensemble at modified sedimentation velocities [15,47]. Only those particles that escape from collision with the side wall, exhibit ideal sedimentation behavior. The latter fraction is larger if the centrifuge tube is placed in a swinging bucket following the direction of the centrifugal field. Further improvement can be achieved by using either radially shaped centrifuge tubes [15] or zonal rotors that avoid wall effects by allowing the sedimentation to proceed within a bowl-shaped chamber [32].

Notwithstanding these uncertainties, the present results have shown that close agreement with sedimentation coefficients, either experimentally measured by DCS or predicted by theoretical modeling, can be achieved. Consequently, the three rather different approaches in determining sedimentation coefficients complement each other very well. Sedimentation coefficient distributions, which are readily accessible by DCS, have proven a valuable tool to optimize the centrifugal separation of nanoparticle mixtures.

4. Conclusions

The agglomeration of nanoparticles yields a large variety of supraparticles, which differ in their aggregation numbers, their compositions, their spatial dimensions, and, finally, their shapes. Clusters of a limited number of nanoscale constituents underlie Brownian motion and have well-defined geometries if prepared by a template-based assembling strategy. This makes the latter ideal model systems for testing analytical tools that are suitable for exploring multimodal mixtures of complex nanoparticles. The latter is challenging and time-consuming with common analytical methods. This is because they provide either poor statistics, such as electron microscopy, or require separation into monomodal fractions prior to their use. The strength of DCS to investigate multimodal mixtures lies in the fact that DCS performs the analytics online, while the nanoparticle mixture is being separated into its individual components. The application of DCS was demonstrated by the analysis of mixtures of colloidal clusters. Nanoparticle clusters of more than 12 constituent particles could be resolved as discrete bands. This, in turn, allowed for an immediate allocation of the observed bands to distinct species. Without any sophisticated mathematical deconvolution, DCS gave access to accurate sedimentation coefficient distributions, simply by using the elementary units of the assemblies as an internal reference. In this way, DCS facilitates an easy-to-use and efficient analysis of colloids with anisotropic shapes, which do not obey Stokes' law. This was verified by comparison with theoretical predictions of sedimentation coefficients for various species of the multimodal mixture by hydrodynamic bead-shell modeling, which considers the exact geometry of the species.

The sedimentation coefficient distributions determined by DCS can be used to optimize routines for preparative centrifugal separations. This study has shown the separation of a total of 11 different populations of colloidal molecules into defined zones within a sucrose density gradient. In addition, the estimation of sedimentation coefficients from the locations of the zones was demonstrated. In this context, the present work could extend an approach, originally established in biological separations, to synthetic nanoparticles.

The feasibility of DCS analysis and its practical application to optimize centrifugal separations suggests the broad applicability of DCS to other nanoparticle systems, including irregularly shaped colloids or mixtures of compositionally heterogeneous nanoparticles.

Supplementary Materials: The following are available online at https://www.mdpi.com/article/10.3390/nano11041027/s1, Figure S1: DCS raw data: light extinction recorded as the function of the sedimentation time, Figure S2: Centrifugal run log to estimate the effective time of centrifugation, Figure S3: Profile of the sucrose density gradient built in the centrifuge tube, Figure S4: Viscosities of aqueous sucrose solutions as functions of temperature and concentration; Figure S5: Densities of aqueous sucrose solutions as functions of temperature and concentration.

Author Contributions: Conceptualization, C.S.P. and A.W.; sedimentation analysis, C.S.P.; hydrodynamic modeling, R.S.; data validation, C.S.P., R.S. and A.W.; investigation, C.S.P. and R.S.; writing—original draft preparation, A.W.; writing—review and editing, C.S.P., R.S. and A.W.; visualization and figures, C.S.P.; supervision, A.W.; project administration, C.S.P.; funding acquisition, A.W. All authors have read and agreed to the published version of the manuscript.

Funding: This research was funded by the DEUTSCHE FORSCHUNGSGEMEINSCHAFT within SFB1214/B4.

Data Availability Statement: The datasets generated during and/or analyzed during the current study are available from the corresponding author on reasonable request.

Acknowledgments: Technical support from the Particle Analysis Centre (PAC) and the Nanostructure Laboratory (nano.lab) at the University of Konstanz is gratefully acknowledged.

Conflicts of Interest: The authors declare no conflict of interest. The funders had no role in the design of the study; in the collection, analyses, or interpretation of data; in the writing of the manuscript, or in the decision to publish the results.

References

1. Everett, D.H. Manual of Symbols and Terminology for Physicochemical Quantities and Units, Appendix II: Definitions, Terminology and Symbols in Colloid and Surface Chemistry. *Pure Appl. Chem.* **1972**, *31*, 577–638. [CrossRef]
2. International Organization for Standardization. *Nanotechnologies Vocabulary Part 2: Nano-Objects (ISO/TS 80004-2:2015)*; International Organization for Standardization: Vernier, Switzerland, 2015.
3. Petosa, A.R.; Jaisi, D.P.; Quevedo, I.R.; Elimelech, M.; Tufenkji, N. Aggregation and Deposition of Engineered Nanomaterials in Aquatic Environments: Role of Physicochemical Interactions. *Environ. Sci. Technol.* **2010**, *44*, 6532–6549. [CrossRef]
4. Israelachvili, J.N. *Intermolecular and Surface Forces*, 3rd ed.; Academic Press: San Diego, CA, USA, 2011; pp. 1–674. [CrossRef]
5. Aimable, A.; Delomenie, A.; Cerbelaud, M.; Videcoq, A.; Chartier, T.; Boutenel, F.; Cutard, T.; Dusserre, G. An experimental and simulation study of heteroaggregation in a binary mixture of alumina and silica colloids. *Colloids Surf. A* **2020**, *605*, 125350. [CrossRef]
6. Trefalt, G.; Cao, T.C.; Sugimoto, T.; Borkovec, M. Heteroaggregation between Charged and Neutral Particles. *Langmuir* **2020**, *36*, 5303–5311. [CrossRef] [PubMed]

7.	García, A.G.; Nagelkerke, M.M.B.; Tuinier, R.; Vis, M. Polymer-mediated colloidal stability: On the transition between adsorption and depletion. *Adv. Colloid Interface Sci.* **2020**, *275*, 102077. [CrossRef] [PubMed]

8.	Anderson, W.; Kozak, D.; Coleman, V.A.; Jämting, Å.K.; Trau, M. A comparative study of submicron particle sizing platforms: Accuracy, precision and resolution analysis of polydisperse particle size distributions. *J. Colloid Interface Sci.* **2013**, *405*, 322–330. [CrossRef]

9.	Holthoff, H.; Borkovec, M.; Schurtenberger, P. Determination of light-scattering form factors of latex particle dimers with simultaneous static and dynamic light scattering in an aggregating suspension. *Phys. Rev. E* **1997**, *56*, 6945–6953. [CrossRef]

10.	Wagner, C.S.; Shehata, S.; Henzler, K.; Yuan, J.Y.; Wittemann, A. Towards nanoscale composite particles of dual complexity. *J. Colloid Interface Sci.* **2011**, *355*, 115–123. [CrossRef]

11.	*Analytical Ultracentrifugation: Techniques and Methods*; Scott, D.J., Harding, S.E., Rowe, A.J., Eds.; Royal Society of Chemistry: Cambridge, UK, 2005; pp. 1–587. [CrossRef]

12.	Laidlaw, I.; Steinmetz, M. Introduction to Differential Sedimentation. In *Analytical Ultracentrifugation: Techniques and Methods*; Scott, D.J., Harding, S.E., Rowe, A.J., Eds.; Royal Society of Chemistry: Cambridge, UK, 2005; pp. 270–290. [CrossRef]

13.	Kamiti, M.; Boldridge, D.; Ndoping, L.M.; Remsen, E.E. Simultaneous Absolute Determination of Particle Size and Effective Density of Submicron Colloids by Disc Centrifuge Photosedimentometry. *Anal. Chem.* **2012**, *84*, 10526–10530. [CrossRef]

14.	Fielding, L.A.; Mykhaylyk, O.O.; Armes, S.P.; Fowler, P.W.; Mittal, V.; Fitzpatrick, S. Correcting for a Density Distribution: Particle Size Analysis of Core–Shell Nanocomposite Particles Using Disk Centrifuge Photosedimentometry. *Langmuir* **2012**, *28*, 2536–2544. [CrossRef]

15.	Schumaker, V.N. Zone Centrifugation. In *Advances in Biological and Medical Physics*; Lawrence, J.H., Gofman, J.W., Eds.; Academic Press: New York, NY, USA, 1967; Volume 11, pp. 245–339. [CrossRef]

16.	Stuckert, R.; Plüisch, C.S.; Wittemann, A. Experimental Assessment and Model Validation on How Shape Determines Sedimentation and Diffusion of Colloidal Particles. *Langmuir* **2018**, *34*, 13339–13351. [CrossRef]

17.	Minelli, C.; Sikora, A.; Garcia-Diez, R.; Sparnacci, K.; Gollwitzer, C.; Krumrey, M.; Shard, A.G. Measuring the size and density of nanoparticles by centrifugal sedimentation and flotation. *Anal. Methods* **2018**, *10*, 1725–1732. [CrossRef]

18.	Mahl, D.; Diendorf, J.; Meyer-Zaika, W.; Epple, M. Possibilities and limitations of different analytical methods for the size determination of a bimodal dispersion of metallic nanoparticles. *Colloids Surf. A* **2011**, *377*, 386–392. [CrossRef]

19.	Wang, R.; Ji, Y.; Wu, X.; Liu, R.; Chen, L.; Ge, G. Experimental determination and analysis of gold nanorod settlement by differential centrifugal sedimentation. *RSC Adv.* **2016**, *6*, 43496–43500. [CrossRef]

20.	Nadler, M.; Mahrholz, T.; Riedel, U.; Schilde, C.; Kwade, A. Preparation of colloidal carbon nanotube dispersions and their characterisation using a disc centrifuge. *Carbon* **2008**, *46*, 1384–1392. [CrossRef]

21.	Arosio, P.; Cedervall, T.; Knowles, T.P.J.; Linse, S. Analysis of the length distribution of amyloid fibrils by centrifugal sedimentation. *Anal. Biochem.* **2016**, *504*, 7–13. [CrossRef] [PubMed]

22.	Neumann, A.; Hoyer, W.; Wolff, M.W.; Reichl, U.; Pfitzner, A.; Roth, B. New method for density determination of nanoparticles using a CPS disc centrifuge™. *Coll. Surf. B* **2013**, *104*, 27–31. [CrossRef]

23.	Akpinar, B.; Fielding, L.A.; Cunningham, V.J.; Ning, Y.; Mykhaylyk, O.O.; Fowler, P.W.; Armes, S.P. Determining the Effective Density and Stabilizer Layer Thickness of Sterically Stabilized Nanoparticles. *Macromolecules* **2016**, *49*, 5160–5171. [CrossRef]

24.	Monopoli, M.P.; Walczyk, D.; Campbell, A.; Elia, G.; Lynch, I.; Baldelli Bombelli, F.; Dawson, K.A. Physical−Chemical Aspects of Protein Corona: Relevance to in Vitro and in Vivo Biological Impacts of Nanoparticles. *J. Am. Chem. Soc.* **2011**, *133*, 2525–2534. [CrossRef]

25.	Davidson, A.M.; Brust, M.; Cooper, D.L.; Volk, M. Sensitive Analysis of Protein Adsorption to Colloidal Gold by Differential Centrifugal Sedimentation. *Anal. Chem.* **2017**, *89*, 6807–6814. [CrossRef]

26.	Van Blaaderen, A. Materials science—Colloids get complex. *Nature* **2006**, *439*, 545–546. [CrossRef] [PubMed]

27.	Li, W.Y.; Palis, H.; Merindol, R.; Majimel, J.; Ravaine, S.; Duguet, E. Colloidal molecules and patchy particles: Complementary concepts, synthesis and self-assembly. *Chem. Soc. Rev.* **2020**, *49*, 1955–1976. [CrossRef]

28.	Plüisch, C.S.; Wittemann, A. Assembly of Nanoparticles into "Colloidal Molecules": Toward Complex and yet Defined Colloids with Exciting Perspectives. In *Advances in Colloid Science*; Rahman, M.M., Asiri, A.M., Eds.; InTech Europe: Rijeka, Croatia, 2016; pp. 237–264. [CrossRef]

29.	Wagner, C.S.; Lu, Y.; Wittemann, A. Preparation of Submicrometer-Sized Clusters from Polymer Spheres Using Ultrasonication. *Langmuir* **2008**, *24*, 12126–12128. [CrossRef] [PubMed]

30.	Manoharan, V.N.; Elsesser, M.T.; Pine, D.J. Dense Packing and Symmetry in Small Clusters of Microspheres. *Science* **2003**, *301*, 483. [CrossRef] [PubMed]

31.	Schwarz, I.; Fortini, A.; Wagner, C.S.; Wittemann, A.; Schmidt, M. Monte Carlo computer simulations and electron microscopy of colloidal cluster formation via emulsion droplet evaporation. *J. Chem. Phys.* **2011**, *135*, 244501. [CrossRef] [PubMed]

32.	Plüisch, C.S.; Bössenecker, B.; Dobler, L.; Wittemann, A. Zonal rotor centrifugation revisited: New horizons in sorting nanoparticles. *RSC Adv.* **2019**, *9*, 27549–27559. [CrossRef]

33.	Sun, X.; Tabakman, S.M.; Seo, W.-S.; Zhang, L.; Zhang, G.; Sherlock, S.; Bai, L.; Dai, H. Separation of Nanoparticles in a Density Gradient: FeCo@C and Gold Nanocrystals. *Angew. Chem. Int. Ed.* **2009**, *48*, 939–942. [CrossRef] [PubMed]

34. Sweetlove, L.J.; Taylor, N.L.; Leaver, C.J. Isolation of Intact, Functional Mitochondria from the Model Plant Arabidopsis thaliana. In *Mitochondria: Practical Protocols*; Leister, D., Herrmann, J.M., Eds.; Humana Press: Totowa, NJ, USA, 2007; pp. 125–136. [CrossRef]

35. Brandrup, J.; Immergut, E.H.; Grulke, E.A. *Polymer Handbook*, 4th ed.; John Wiley & Sons, Inc.: New York, NY, USA, 1999; pp. 1–2336.

36. Kahler, H.; Lloyd, B.J. Density of Polystyrene Latex by a Centrifugal Method. *Science* **1951**, *114*, 34–35. [CrossRef] [PubMed]

37. Yannas, J.B. Highly precise density determination for polymers in latex form. *J. Polym. Sci. Part B Polym. Lett.* **1964**, *2*, 1005–1008. [CrossRef]

38. Hoffmann, M.; Wagner, C.S.; Harnau, L.; Wittemann, A. 3D Brownian Diffusion of Submicron-Sized Particle Clusters. *ACS Nano* **2009**, *3*, 3326–3334. [CrossRef] [PubMed]

39. Lauga, E.; Brenner, M.P. Evaporation-Driven Assembly of Colloidal Particles. *Phys. Rev. Lett.* **2004**, *93*, 238301. [CrossRef] [PubMed]

40. Plüisch, C.S.; Wittemann, A. Shape-Tailored Polymer Colloids on the Road to Become Structural Motifs for Hierarchically Organized Materials. *Macromol. Rapid Commun.* **2013**, *34*, 1798–1814. [CrossRef] [PubMed]

41. Price, C.A. Appendix C—Properties of Gradient Materials. In *Centrifugation in Density Gradients*; Price, C.A., Ed.; Academic Press: London, UK, 1982; pp. 327–389. [CrossRef]

42. Bohren, C.F.; Huffman, D.R. Absorption and Scattering by a Sphere. In *Absorption and Scattering of Light by Small Particles*; Bohren, C.F., Huffman, D.R., Eds.; Wiley-VCH Verlag GmbH & Co. KGaA: Weinheim, Germany, 1998; pp. 82–129. [CrossRef]

43. Scott, D.J.; Schuck, P. A Brief Introduction to the Analytical Ultracentrifugation of Proteins for Beginners. In *Analytical Ultracentrifugation: Techniques and Methods*; Scott, D.J., Rowe, A.J., Eds.; Royal Society of Chemistry: Cambridge, UK, 2005; pp. 1–25. [CrossRef]

44. Spragg, S.P. The Bases of Centrifugal Separations. In *Centrifugal Separations in Molecular and Cell Biology*; Birnie, G.D., Rickwood, D., Eds.; Butterworth-Heinemann: Oxford, UK, 1978; pp. 7–32. [CrossRef]

45. Martin, R.G.; Ames, B.N. A Method for Determining the Sedimentation Behavior of Enzymes: Application to Protein Mixtures. *J. Biol. Chem.* **1961**, *236*, 1372–1379. [CrossRef]

46. Hinton, R.; Dobrota, M. Laboratory Techniques in Biochemistry and Molecular Biology. In *Laboratory Techniques in Biochemistry and Molecular Biology*; Work, T.S., Work, E., Eds.; North-Holland Publishing Company: Oxford, UK, 1978; Volume 6, pp. 243–253. [CrossRef]

47. Price, C.A. Sedimentation Theory: A Semiquantitative Approach. In *Centrifugation in Density Gradients*; Price, C.A., Ed.; Academic Press: London, UK, 1982; pp. 32–78. [CrossRef]

 *nanomaterials*

Article

Multimaterial 3D Printing for Arbitrary Distribution with Nanoscale Resolution

Fengqiang Zhang [1], Changhai Li [2,*], Zhenlong Wang [1,2,*], Jia Zhang [1,2] and Yukui Wang [1]

[1]   School of Mechatronics Engineering, Harbin Institute of Technology, Harbin 150001, China

[2]   Key Laboratory of Micro-systems and Micro-structures Manufacturing, Ministry of Education, Harbin Institute of Technology, Harbin 150080, China

*   Correspondence: lich@hit.edu.cn (C.L.); wangzl@hit.edu.cn (Z.W.); Tel.: +86-451-8641-3485 (Z.W.)

Received: 3 July 2019; Accepted: 30 July 2019; Published: 2 August 2019

Abstract: At the core of additive manufacturing (3D printing) is the ability to rapidly print with multiple materials for arbitrary distribution with high resolution, which can remove challenges and limits of traditional assembly and enable us to make increasingly complex objects, especially exciting meta-materials. Here we demonstrate a simple and effective strategy to achieve nano-resolution printing of multiple materials for arbitrary distribution via layer-by-layer deposition on a special deposition surface. The established physical model reveals that complex distribution on a section can be achieved by vertical deformation of simple lamination of multiple materials. The deformation is controlled by a special surface of the mold and a contour-by-contour (instead of point-by-point) printing mode is revealed in the actual process. A large-scale concentric ring array with a minimum feature size below 50 nm is printed within less than two hours, verifying the capacity of high-throughput, high-resolution and rapidity of printing. The proposed printing method opens the way towards the programming of internal compositions of object (such as functional microdevices with multiple materials).

Keywords: 3D printing; nano-resolution; arbitrary distribution; multimaterials; deposition surface; rapidity; large scale

---

1. Introduction

Three-dimensional printing is well-known as one of the disruptive technologies affecting the future, which may bring about the third industrial revolution. Due to its capacity of rapid prototyping for arbitrary shape, it is widely applied in aerospace [1], medical health [2,3], structural electronics [4], and mold [5]. The traditional 3D printing technologies including stereolithography [6,7], selective laser sintering [8,9], fused deposition modeling [10,11] have consistently been improved, however they are still subject to the low-throughput and low-resolution fundamentally resulting from point-by-point printing and large printing unit size, in which the optimum printing resolution of approximately 20–50 μm is available [12]. Therefore, as an extension of 3D printing, continuous liquid interface production is developed with higher efficiency [13–15], as well as femtosecond laser direct writing based on two-photon polymerization with higher resolution of up to 100 nm [16–21]. Nevertheless, a trade-off of efficiency and resolution is still difficult to realize.

In addition, without an effective and precise controlling strategy of position of multiple materials, the state-of-the-art 3D printing technology still faces enormous challenges to achieve real printing of multiple materials for arbitrary distribution [22–28]. The study of 3D printing has been stagnating at the primary stage of controlling object shape for a long time [6] and a serious breakthrough is needed for the senior stage of programming internal compositions [29], despite some achievements such as the printing with hybrid multimaterials [22], printing via multiple nozzles with different materials [23,30], simple and imprecise controlling of compositions by the assisted physical morphology [24–26], magnetism [27], chemical reaction [28], etc. Three-dimensional printing of multiple materials with nano-resolution can create further applications in micro- and nano-scale multifuctional device [31,32] and high-level biological tissues [33]. Therefore, how to control multiple materials with high resolution and efficiency to construct objects has attracted more interest from the researchers. As mentioned by American scientist Hod Lipson, the core of real 3D printing was the capacity of achieving arbitrary distribution of multiple materials [29], which is what we have been doing.

In this study, we demonstrate the precise control of the position of multiple materials for arbitrary distribution by means of the morphological characteristics. The established physical model shows that a complex distribution on a section can be achieved by the vertical deformation of multimaterial lamination. The deformation is controlled by a special surface of the mold and a contour-by-contour printing mode is revealed in the actual process. A large-scale concentric ring array with a minimum feature size below 50 nm is printed within less than two hours, demonstrating the nano-resolution and rapidity of printing. Simultaneous improvement of resolution and efficiency considered as a pair of contradictions is realized. The proposed printing strategy makes a great extension of function and application of 3D printing technologies.

2. Experimental Details

2.1. Fabrication of Concentric Ring Array

A 2-inch patterned sapphire substrate (PSS, Zhejiang Bolant Semiconductor Technology Co., Ltd., Zhejiang, China.) with conical structure array (Figure S1) was used as the deposition substrate and cleaned by argon (Ar) plasma before deposition of materials. Then the alternative depositions of tungsten and aluminium (99.99% purity, Beijing Goodwill Metal, Co., Ltd., Beijing, China) on the PSS were implemented in the argon atmosphere of 0.45 Pa using a high vacuum multiple target magnetron sputtering coating machine (JCP-350M2, Beijing Technol Science Co., Ltd., Beijing, China). Deposition thickness of W and Al was set at 120 nm except the first layer with thickness of 240 nm, and the total number of deposition layers was twenty-three (See Table S1 for details). To achieve uniform thickness of deposition film, the substrate holder was simultaneously cooled with circular liquid nitrogen and rotated during deposition. Finally, the required lamination was obtained (Figure S2) and the concentric ring array on the section $z = 1.83$ μm of lamination was further revealed after removing the upside materials using ultra-precision lathing and polishing paste (Al_2O_3, 10–50 nm, Microspectrum Technology Co., Ltd., Shanghai, China.). The whole process took less than 2 h.

2.2. Characterization

The scanning electron microscope (SEM) images were collected from the field emission scanning electron microscope (Supra 55 Sapphire, Carl Zeiss, Oberkochen, Germany) at an acceleration voltage of 20 kV. The morphology of microstructures array was measured by the atomic force microscope (AFM, dimension fastscan, Bruker, Billerica, MA, USA). The transmission electron microscopy (TEM) specimen was prepared by using Helios NanoLab 600i FIB/SEM Dual-Beam system (FEI, Hillsboro, OR, USA). The TEM, high-resolution TEM (HRTEM), energy-dispersive X-ray (EDX) and high-angle annular dark field-scanning transmission electron microscopy (HAADF-STEM) images were taken from in-situ multifunctional transmission electron microscopy (Talos F200X, FEI, Hillsboro, OR, USA) at an acceleration voltage of 200 kV.

3. Results and Discussion

3.1. 3D Printing Strategy for an Arbitrary Distribution of Multiple Materials

Figure 1 illustrates the 3D printing strategy for an arbitrary distribution of multiple materials. The principle diagram of achieving arbitrary distribution is shown in Figure 1a–d. The lamination S is composed of materials $W(i)$ ($i = 1 \ldots n$) in a certain order, in which an arbitrary curved surface

$z = f(x, y)$ and datum $z = g$ are given (Figure 1a). Deformation from the surface $z = f(x, y)$ to the datum $z = g$ is implemented (Figure 1b). For instance, these points I–VII on the surface $z = f(x, y)$ move vertically h_I–h_{VII} to the datum, respectively. Correspondingly, the transformation from the lamination S to lamination V is forced as well as the bottom plane $z = k$ to the surface $z = F(x, y)$ by the same rule. The distribution of multiple materials is obtained by intercepting the section $z = g$ in the lamination V (Figure 1c). In addition, the curved surface $z = f(x, y)$ is endowed with corresponding materials and projected vertically onto a plane, resulting in the transverse distribution (Figure 1d). The comparison reveals that the distribution on the section $z = g$ is absolutely the same with the transverse distribution of the curved surface $z = f(x, y)$ within the lamination S. Therefore, an arbitrary and complex distribution on a plane can be programmed and achieved via the designing of corresponding curved surface $z = f(x, y)$ and longitudinal deforming of lamination S of multiple materials. The core of the strategy lies in the realizing of transformation from simply longitudinal laminating of multiple materials to complex transverse distribution via longitudinal deformation.

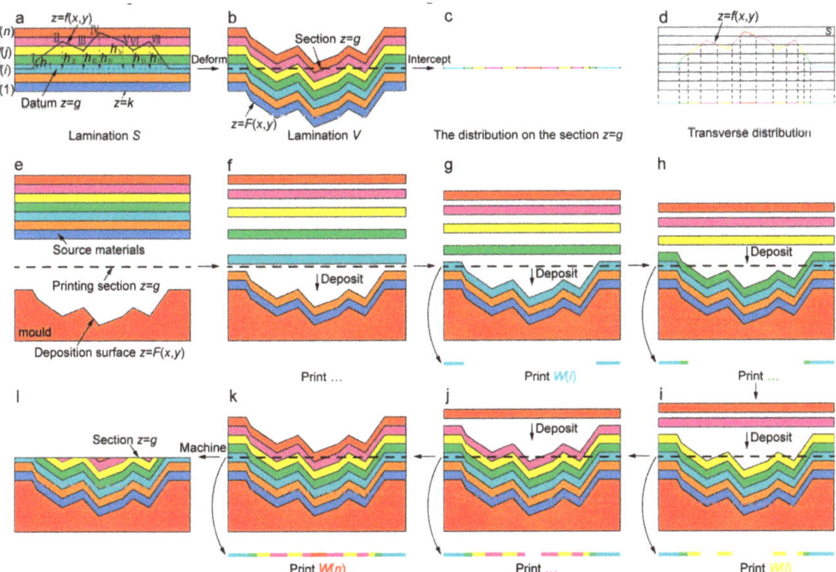

Figure 1. Three-dimensional printing strategy for arbitrary distribution of multiple materials. (**a–d**) Principle diagram of achieving arbitrary distribution of multiple materials. Arbitrary curved surface $z = f(x, y)$ and datum $z = g$ are given within lamination S of multiple materials $W(i)$ (**a**). The deformation (**b**) from the surface $z = f(x, y)$ to the datum $z = g$ is implemented. For instance, the points I–VII of surface $z = f(x, y)$ move vertically h_I–h_{VII} to the datum, respectively. Correspondingly, the transformation from lamination S to lamination V is forced by the same rule. Then the section $z = g$ (**c**) is intercepted from lamination V, on which the multimaterial distribution can be obtained. That the distribution on the section $z = g$ is exactly the same with transverse distribution of surface $z = f(x, y)$ (**d**) is observed. (**e–l**) Schematic diagram of 3D printing process for arbitrary distribution of multiple materials. The deposition surface of the mold is prepared and source materials $W(i)$ ($i = 1 \ldots n$) are designed for printing (**e**). Materials $W(1) \ldots W(i) \ldots W(n)$ are successively deposited on the deposition surface of the mold to further print corresponding material on the section $z = g$ (**f–k**). The section $z = g$ can be obtained by machining deposition body (**l**).

Based on the principle, the schematic diagram of 3D printing process for arbitrary distribution of multiple materials is revealed, in which the longitudinal deformation is controlled by the morphology (Figure 1e–l). According to the bottom surface $z = F(x, y)$ of lamination V (Figure 1b), the deposition surface $z = F(x, y)$ of the mold is prepared and source materials is given for 3D printing (Figure 1e). Then the deposition on the deposition surface is conducted in order of $W(1) \ldots W(i) \ldots W(n)$ through atomic deposition technology such as magnetron sputtering (Figure 1f–k, top), resulting in controllable printing of corresponding material on the section $z = g$ (Figure 1f–k, bottom) and the lamination V can be achieved (Figure 1k). Finally, the designed distribution of multiple materials is obtained by removing these materials upon the section $z = g$ (Figure 1l). Notably, deposition on the deposition surface is obviously the process of printing multiple materials for the distribution on the section $z = g$, and the deposition rate determines the printing speed. Contour-by-contour printing of materials is conducted according to these section contours of deposition surface from top to bottom (Figure 1f–k, bottom), in which the dependence of printing direction is on the height of morphology. The contour-by-contour (rather than point-by-point) printing brings about the rapidity of 3D printing of multiple materials.

Figure 2 illustrates the analysis of the printing resolution. The lamination V is achieved via multimaterial deposition on a conical structure with triangular contour, in which the slope of the mold and the thickness of deposition layer $W(i)$ are α and h, respectively (Figure 2a). Then the width w of $W(i)$ on the section $z = g$ in the lamination V is investigated. There is an equation $w = h/\tan(\alpha)$ resulting in the width w of $W(i)$ decreasing with the decrease of the thickness h of $W(i)$ or the increase of the slope α of the mold. The width w_1 and w_2 of $W(i)$ on the section $z = g$ are obtained by decreasing the thickness ($h_1 < h$) of deposition layer $W(i)$ and increasing the slope ($\alpha_2 > \alpha$) of the mold, resulting in the inequalities $w_1 < w$ and $w_2 < w$, respectively (Figure 2b,c). From Figure 2, the width w of $W(i)$ on the section $z = g$ can theoretically reach nanoscale and even atomic level, when the thickness h of $W(i)$ is small enough or the slope α of the mold is lager enough. Notably, the thickness of the monolayer atoms can be easily fabricated by the atomic deposition. Therefore, nanoscale resolution printing is feasible and suitable for the substrate with a complex structure. At this time, a good trade-off between the resolution and throughput is realized.

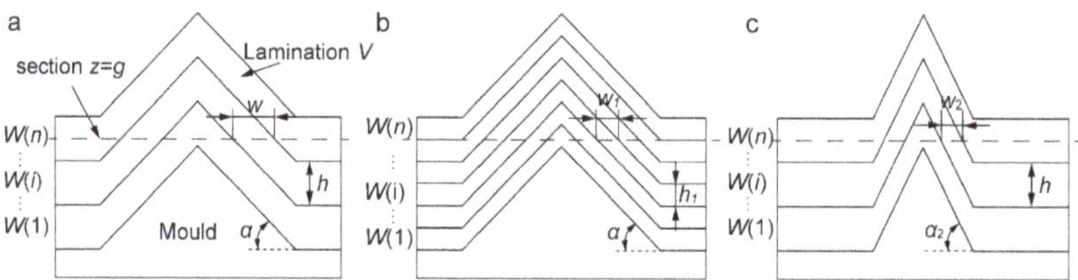

Figure 2. Analysis of the printing resolution. (**a**) The width w of $W(i)$ on the section $z = g$ in the lamination V under the conditions of the slope α of the mold and the thickness h of deposition layer $W(i)$. There is an equation $w = h/\tan(\alpha)$ resulting in the width w of $W(i)$ decreasing with the decrease of the thickness of $W(i)$ or the increase of the slope of the mold. (**b**) The width w_1 of $W(i)$ on the section $z = g$ by decreasing the thickness ($h_1 < h$) of deposition layer $W(i)$. There is an inequality $w_1 < w$. (**c**) The width w_2 of $W(i)$ on the section $z = g$ by increasing the slope ($\alpha_2 > \alpha$) of the mold. There is an inequality $w_2 < w$.

In the process of 3D printing, the deposition surface acts as a series of masks, and contour-by-contour printing of same-material on the section $z = g$ is conducted to further achieve self-assembly of multiple materials. The distribution on the section $z = g$ is actually a mapping of the deposition surface and mapped by laminating multiple materials on the deposition surface. Therefore, this is a new method of 3D printing for rapidly achieving an arbitrary nano-resolution distribution of multiple materials on a large scale, which is called Morphology-Guided Printing (MGP).

3.2. Concentric Ring Array Fabricated by MGP Method

Figure 3a–f schematically illustrates the MGP process of multiple materials on the printing section $z=g$ based on the PSS with conical structure array (Figure S1). Half of a single conical structure is used for observing and has a like-parabola contour. The pattern on the section $z = g$ is printed circle by circle from center to edge via the deposition of multiple materials. Using the MGP method, a concentric ring array is fabricated via alternative deposition of tungsten and aluminium within less than two hours for investigation (Figure 3g–j and Table S1). Figure 3g,h presents top and side view SEM images of pre-processing multimaterial lamination with a thickness of 4.32 μm, respectively. Honeycomb contour and good uniformity of microstructures (Figure S2) of lamination are observed, and uniformity of thickness of deposition layer and alternative laminating of tungsten (white) and aluminium (black) are demonstrated by the internal structure of lamination (red dotted box in Figure 3h). This demonstrates good capacity of deposition of materials and provides strong support for high-solution printing of multiple materials on a certain section. However, some bubbles on the lamination are observed, which may be the result of atomic aggregation in the process of deposition and affect printing accuracy of materials. Figure 3i,j reveals top and side SEM images of post-processing lamination of materials, respectively. The section $z = 1.83$ μm is obtained via removing upper materials, and the uniform concentric ring array is further developed. However, the insert of Figure 3i shows tortuous boundary of rings, resulting from these bubbles. For the section $z = 1.83$ μm of single microstructure, sequential same-material rings are printed circle by circle from center to edge and at the center of the section is just aluminum (red dotted circle shown in Figure 3j). Generally, a large-scale distribution of multiple materials can be rapidly printed on the specified section.

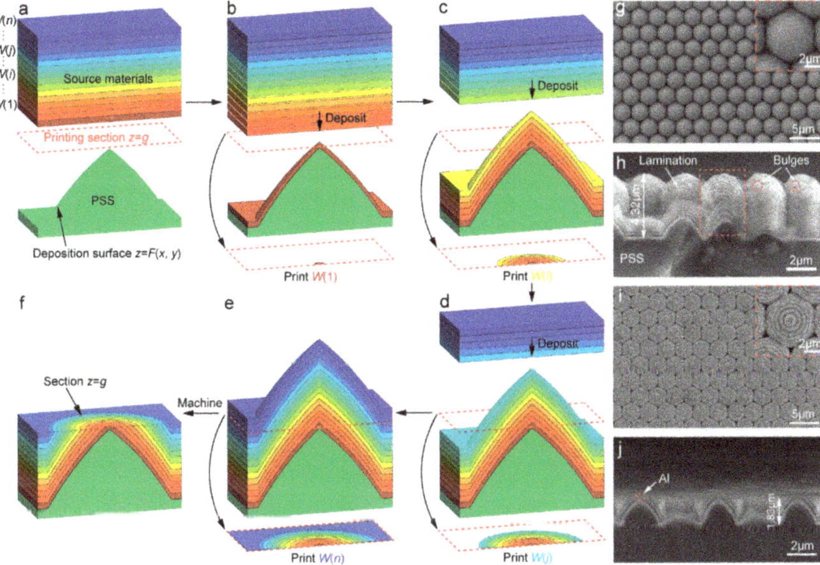

Figure 3. Morphology-Guided Printing (MGP) of multiple materials on printing section $z = g$ based on patterned sapphire substrate (PSS). (**a–f**) Schematic diagram of MGP process. PSS and source materials $W(i)$ ($i = 1 \ldots n$) for printing are given (**a**). Materials $W(1) \ldots W(i) \ldots W(n)$ are successively printed circle by circle on the section $z = g$ (**b–e**). The section $z = g$ is obtained by machining (**f**). (**g–j**) MGP of concentric circle array. Top (**g**) and side (**h**) view SEM images of pre-processing multimaterial lamination with a thickness of 4.32 μm are revealed on the PSS. Lamination is formed via alternative depositing of tungsten and aluminum. Honeycomb contour of lamination and good uniformity of microstructures of lamination is demonstrated in (**g**) and a single microstructure is revealed in the insert. W (white)-and-Al (black) alternating lamination (red dotted box) and bulges on the surface of lamination (red dotted circle) are observed in (**h**). Top (**i**) and side (**j**) view SEM images of post-processing lamination are shown. The section $z = 1.83$ μm is obtained via removing materials. Uniform concentric circle array is shown, and a single concentric circle is revealed in the insert. At the center of the section is just aluminum.

To further investigate the printing capacity of MGP method for arbitrary distribution of multiple materials, the section $z = 1.83$ μm with a thickness of less than 100 nm is intercepted by focused ion beam (FIB) and characterized by TEM (Figure 4a–f). Figure 4b shows HAADF-STEM image of single concentric rings unit (red dotted box of Figure 4a), which demonstrates W (white)-and-Al (black) alternative concentric rings. Figure 4c–e exhibits the EDX elemental mapping images of Al, W and their overlap, respectively, which further verify W-and-Al alternative concentric distribution. Figure 4f reveals Al/W component distribution curves along the purple line shown (shown in Figure 4b), in which the position of zero coordinate presents the center of the concentric ring. The transforming of like-rectangle to triangle wave is accompanied by the decreasing of the period in the radial direction. This indicates the widths of the concentric rings have a decreasing trend with increasing in radius, which is caused by the increasing slope of the conical structure (Figure S1). And the W content of less than 100% indicates the doping in W region. Notably, the doping may be caused by atomic diffusion or environmental pollution in the process of deposition. For quantifying the distribution size of materials, the TEM image of specimen is processed using MATLAB software to get ten concentric nanorings with clear and tortuous boundaries (Figure 4g). Then the mean radii and width of distribution of materials are calculated (Table 1), further verifying the decreasing trend of the ring widths. The emphasis is on that the smallest feature size of 41.8nm demonstrates the nanoscale resolution of MGP method.

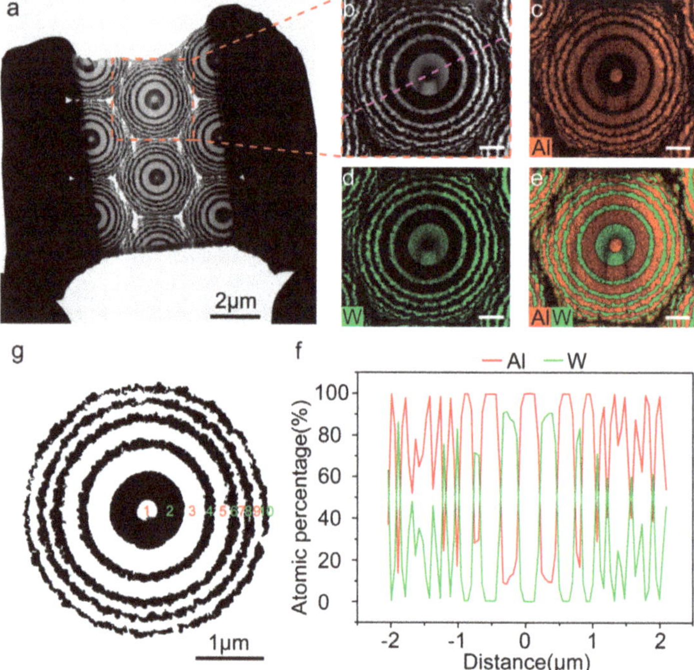

Figure 4. Elemental analysis of single concentric ring unit. (**a**) TEM image of specimen. The section $z = 1.83$ μm with W-and-Al alternating rings printed is intercepted by FIB and the thickness of specimen is less than 100 nm. (**b**) HAADF-STEM image of single concentric ring unit shown in red dotted box of (**a**). EDX elemental mapping images of Al (**c**), W (**d**), and their overlap (**e**) are shown. (**f**) Material distribution curves along the purple dotted line shown in (**b**). Scale bar: 500 nm in (**b–e**). (**g**) Post-processing material distribution including ten concentric rings using MATLAB software. Material distribution is extracted from TEM image of specimen (**a**) using image processing technology and clearer boundaries are shown. Al (white)-and-W (black) alternative concentric rings are labeled 1 to 10 in turn.

Table 1. Radius and width of distribution of multiple materials in Figure 4g.

Concentric rings [a]	1	2	3	4	5	6	7	8	9	10
Circle radius [b] (nm)	65.7 ± 9.5	228.3 ± 8.6	372.3 ± 12.4	424.9 ± 11.4	524.4 ± 16.5	578.2 ± 14.7	623.9 ± 16.0	668.4 ± 12.7	710.6 ± 14.4	752.4 ± 13.7
Ring width [c] (nm)	65.7	162.6	144	52.6	99.5	53.8	45.7	44.5	42.2	41.8

[a] The concentric rings marked 1 to 10 are shown in Figure 4g. [b] Presenting the mean radius and error of outer circle of each ring calculated by MATLAB software. [c] Determined by the difference between outer and inner radius of each ring.

Then the crystal analysis on the section $z = 1.83$ μm is conducted (Figure 5). Figure 5a shows the TEM image of specimen, where the white and black regions represent the materials of Al and W, respectively. Then the corresponding HRTEM images of these regions marked with red boxes shown in Figure 5a and their selected area electron diffraction (SAED) patterns are exhibited in Figure 5b–i top and bottom, respectively. The evolution of crystal state is shown in Figure 5b–e, in which Figure 5b displays the good single crystal nature of Al in the pure Al region, the crystals of W begin to appear with the doping of W in the transition region (Figure 5c,d) and Figure 5e exhibits the amorphous nature of W in pure W region. Additionally, the red curves (shown in Figure 5c,d) are the boundaries between pure Al, transition and pure W regions. It is worth noting that the single crystals nature of Al is not destroyed by a small amount of doped W in transition region (Figure 5c,d). However, the poly-crystallization occurs in other regions such as transition (Figure 5f,h), pure Al (Figure 5g), and W (Figure 5i) regions, induced by the doping of other elements. This further results in refinement of crystals, low crystallinity and diversification of crystal orientation. Generally, a small amount of doping in the process of printing is further revealed by the crystal analysis, and the same-material printing region of small enough size is vulnerable to contamination. Meanwhile, the doping can be eliminated by some measures, such as using the targets of higher purity, prevention of mutual contamination of targets, improved cooling system for restraining atomic diffusion, and maintaining the adequate cleanliness of the sputtering chamber by cleaning up impurities and an improved vacuum system.

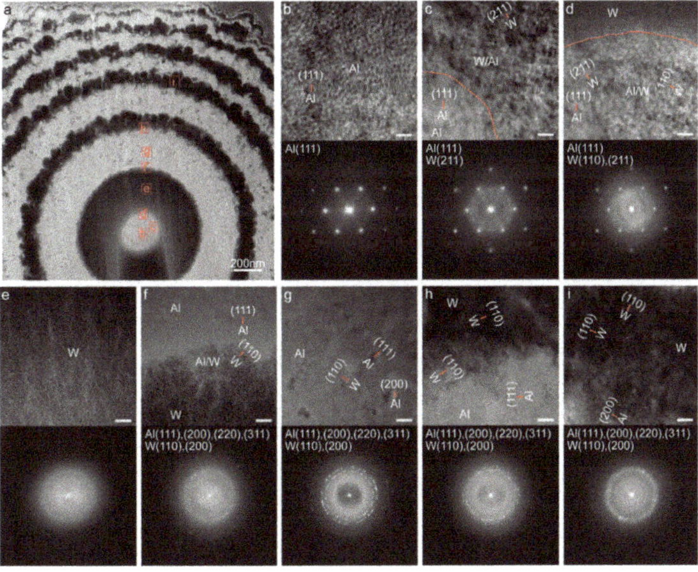

Figure 5. Crystal analysis of specimen. (**a**) TEM image of specimen. (**b–i**) HRTEM images (top) and corresponding SAED patterns (bottom) of these regions marked with red boxes in (**a**). Scale bar: 10 nm in (**b–i**).

In addition, using the same substrate (such as PSS), other distributions are available by programming the material type, sequence and thickness of each deposition layer. For instance, the radial gradient circle can be printed by depositing gradient materials on the PSS (Figure 3a–f). Furthermore, when the substrate with a different morphology is used as the substrate, the corresponding distribution can be achieved via the design and depositing of multiple materials. Figure S3 exhibits a printed distribution on the specified section via depositing gradient materials on the morphology of monolayer cells. By comparison, a high consistency between SEM image of the distribution and optical image of the morphology is observed. Therefore, the feasibility and high resolution of MGP method is further validated as well as the diversity of distribution of multiple materials.

4. Summary

In summary, a new method of nano-resolution 3D printing for rapidly achieving arbitrary distribution of multiple materials on a large scale is proposed. The designed distribution of multiple materials on the specified section can be obtained by the corresponding deformation of lamination of multiple materials. In fact, the deformation of lamination can be realized via the corresponding morphology. Based on atomic deposition on PSS, a concentric ring array with nanoscale resolution on a large scale is printed verifying the rapidity and nanoscale resolution of MGP method. The improvement of atomic deposition including collimation and secondary deposition, and the decreasing of thickness of each layer, will contribute to higher and even atomic resolution of MGP. In addition, almost all materials can be used and are mutually compatible for MGP due to the abundant deposition methods including physical vapor deposition (PVD) and chemical vapor deposition (CVD). Versatile MGP method is also suitable for particle deposition to print macro-scale architecture. More importantly, these sections with the designed distribution can be isolated by micro–nano processing technology such as FIB and further laminated to form an arbitrary distribution of multiple materials in three-dimensional space. Therefore, MGP method makes a great extension of function and application of additive manufacturing and opens the way towards the programming internal components of objects such as transistors.

Supplementary Materials: The following are available online at http://www.mdpi.com/2079-4991/9/8/1108/s1, Figure S1: Patterned sapphire substrate (PSS), Table S1: Parameters of magnetron sputtering, Figure S2: AFM image of lamination of multimaterials. Good uniformity of microstructures of lamination is demonstrated. Figure S3: Comparison of the morphology of monolayer cells and the distribution of multiple materials.

Author Contributions: C.L. and Z.W. conceived the project. F.Z. carried out the experiments, measurements, and analysis, and finished the draft. J.Z. participated in the measurements and modified the manuscript. Y.W. put forward some amendments.

Funding: The research was funded by Harbin Boshi Automation Co., Ltd. and Weihai Boshi 3D Technology Co., Ltd. This research was supported by National Natural Science Foundation of China (Grant No. 51775145) and the Major Project of Applied Technology Research and Development Plan of Heilongjiang Province (Grant No. GA16A404).

Acknowledgments: We thank Yongze Guo for processing TEM image, Junjie Yang and Likuan Zhu for revision.

Conflicts of Interest: The authors declare no conflict of interest.

References

1. Suresh, S.; Mortensen, A. Functionally graded metals and metal-ceramic composites: Part 2 thermomechanical behaviour. *Int. Mater. Rev.* **1997**, *42*, 85–116. [CrossRef]
2. Rutz, A.L.; Hyland, K.E.; Jakus, A.E.; Burghardt, W.R.; Shah, R.N. A multimaterial bioink method for 3D printing tunable, cell-compatible hydrogels. *Adv. Mater.* **2015**, *27*, 1607–1614. [CrossRef] [PubMed]
3. Han, X.X.; Bibb, R.; Harris, R. Engineering design of artificial vascular junctions for 3D printing. *Biofabrication* **2016**, *8*, 025018. [CrossRef] [PubMed]
4. Macdonald, E.; Salas, R.; Espalin, D.; Perez, M.; Aguilera, E.; Muse, D.; Wicker, R.B. 3D printing for the rapid prototyping of structural electronics. *IEEE Access* **2014**, *2*, 234–242. [CrossRef]

5. Ebendorff-Heidepriem, H.; Schuppich, J.; Dowler, A.; Lima-Marques, L.; Monro, T.M. 3D-printed extrusion dies: A versatile approach to optical material processing. *Opt. Mater. Express* **2014**, *8*, 1494–1504. [CrossRef]

6. Liu, T.; Guessasma, S.; Zhu, J.H.; Zhang, W.H.; Belhabib, S. Functionally graded materials from topology optimisation and stereolithography. *Eur. Polym. J.* **2018**, *108*, 199–211. [CrossRef]

7. Chia, H.N.; Wu, B.M. Recent advances in 3D printing of biomaterials. *J. Biol. Eng.* **2015**, *9*, 4. [CrossRef]

8. Yap, C.Y.; Chua, C.K.; Dong, Z.L.; Liu, Z.H.; Zhang, D.Q.; Loh, L.E.; Sing, S.L. Review of selective laser melting: Materials and applications. *Appl. Phys. Rev.* **2015**, *2*, 041101. [CrossRef]

9. Yan, A.R.; Wang, Z.Y.; Yang, T.T.; Wang, Y.L.; Ma, Z.H. Sintering densification behaviors and microstructural evolvement of W-Cu-Ni composite fabricated by selective laser sintering. *Int. J. Adv. Manuf. Technol.* **2017**, *90*, 657–666. [CrossRef]

10. Gnanasekaran, K.; Heijmans, T.; Van Bennekom, S.; Woldhuis, H.; Wijnia, S.; De With, G.; Friedrich, H. 3D printing of CNT-and graphene-based conductive polymer nanocomposites by fused deposition modeling. *Appl. Mater. Today* **2017**, *9*, 21–28. [CrossRef]

11. Kitson, P.J.; Glatzel, S.; Chen, W.; Lin, C.G.; Song, Y.F.; Cronin, L. 3D printing of versatile reactionware for chemical synthesis. *Nat. Protoc.* **2016**, *11*, 920–936. [CrossRef]

12. Chen, X.F.; Liu, W.Z.; Dong, B.Q.; Lee, J.; Ware, H.O.T.; Zhang, H.F.; Sun, C. High-speed 3D printing of millimeter-size customized aspheric imaging lenses with sub 7 nm surface roughness. *Adv. Mater.* **2018**, *30*, 1705683. [CrossRef]

13. Tumbleston, J.R.; Shirvanyants, D.; Ermoshkin, N.; Janusziewicz, R.; Johnson, A.R.; Kelly, D.; Chen, K.; Pinschmidt, R.; Rolland, J.P.; Ermoshkin, A.; et al. Continuous liquid interface production of 3D objects. *Science* **2015**, *347*, 1349–1352. [CrossRef]

14. Hu, R.; Huang, B.X.; Xue, Z.H.; Li, Q.Y.; Xia, T.; Zhang, W.; Lu, C.H.; Xu, H.G. Synthesis of photocurable cellulose acetate butyrate resin for continuous liquid interface production of three-dimensional objects with excellent mechanical and chemical-resistant properties. *Carbohydr. Polym.* **2019**, *207*, 609–618. [CrossRef]

15. Lin, W.X.; Liu, H.G.; Huang, H.Z.; Huang, J.H.; Ruan, K.M.; Lin, Z.X.; Wu, H.C.; Zhang, Z.; Chen, J.M.; Li, J.H. Enhanced continuous liquid interface production with track-etched membrane. *Rapid Prototyp. J.* **2019**, *25*, 117–125. [CrossRef]

16. Zhao, Y.Y.; Zhang, Y.L.; Zheng, M.L.; Dong, X.Z.; Duan, X.M.; Zhao, Z.S. Three-dimensional Luneburg lens at optical frequencies. *Laser Photon. Rev.* **2016**, *10*, 665–672. [CrossRef]

17. Xu, B.B.; Xia, H.; Niu, L.G.; Zhang, Y.L.; Sun, K.; Chen, Q.D.; Xu, Y.; Lv, Z.Q.; Li, Z.H.; Misawa, H.; et al. Flexible nanowiring of metal on nonplanar substrates by femtosecond-laser-induced electroless plating. *Small* **2010**, *6*, 1762–1766. [CrossRef]

18. Cao, Y.Y.; Takeyasu, N.; Tanaka, T.; Duan, X.M.; Kawata, S. 3D metallic nanostructure fabrication by surfactant-assisted multiphoton-induced reduction. *Small* **2009**, *5*, 1144–1148. [CrossRef]

19. Vyatskikh, A.; Delalande, S.; Kudo, A.; Zhang, X.; Portela, C.M.; Greer, J.R. Additive manufacturing of 3D nano-architected metals. *Nat. Commun.* **2018**, *9*, 593. [CrossRef]

20. Ma, Z.C.; Zhang, Y.L.; Han, B.; Chen, Q.D.; Sun, H.B. Femtosecond-laser direct writing of metallic micro/nanostructures: From fabrication strategies to future applications. *Small Methods* **2018**, *2*, 1700413. [CrossRef]

21. Tang, H.; Lin, X.F.; Feng, Z.; Chen, J.Y.; Gao, J.; Sun, K.; Wang, C.Y.; Lai, P.C.; Xu, X.Y.; Wang, Y.; et al. Experimental two-dimensional quantum walk on a photonic chip. *Sci. Adv.* **2018**, *4*, eaat3174. [CrossRef]

22. Zhang, Y.F.; Zhang, N.B.; Hingorani, H.; Ding, N.Y.; Wang, D.; Yuan, C.; Zhang, B.; Gu, G.Y.; Ge, Q. Fast-response, stiffness-tunable soft actuator by hybrid multimaterial 3D printing. *Adv. Funct. Mater.* **2019**, *29*, 1806698. [CrossRef]

23. Li, F.; Macdonald, N.P.; Guijt, R.M.; Breadmore, M.C. Multimaterial 3D printed fluidic device for measuring pharmaceuticals in biological fluids. *Anal. Chem.* **2019**, *91*, 1758–1763. [CrossRef]

24. Su, M.; Huang, Z.D.; Li, Y.F.; Qian, X.; Li, Z.; Hu, X.T.; Pan, Q.; Li, F.Y.; Li, L.H.; Song, Y.L. A 3D self-shaping strategy for nanoresolution multicomponent architectures. *Adv. Mater.* **2017**, *30*, 1703963. [CrossRef]

25. Li, Y.F.; Su, M.; Li, Z.; Huang, Z.D.; Li, F.Y.; Pan, Q.; Ren, W.J.; Hu, X.T.; Song, Y.L.; Su, M.; et al. Patterned arrays of functional lateral heterostructures via sequential template-directed printing. *Small* **2018**, *14*, 1800792. [CrossRef]

26. Huang, Z.D.; Yang, Q.; Su, M.; Li, Z.; Hu, X.T.; Li, Y.F.; Pan, Q.; Ren, W.J.; Li, F.Y.; Song, Y.L. A general approach for fluid patterning and application in fabricating microdevices. *Adv. Mater.* **2018**, *30*, 1802172. [CrossRef]
27. Kokkinis, D.; Schaffner, M.; Studart André, R. Multimaterial magnetically assisted 3D printing of composite materials. *Nat. Commun.* **2015**, *6*, 8643. [CrossRef]
28. Wang, T.; Handschuh-Wang, S.; Huang, L.; Zhang, L.; Jiang, X.; Kong, T.; Zhang, W.; Lee, C.S.; Zhou, X.C.; Tang, Y.B. Controlling directional liquid motion on micro- and nanocrystalline diamond/β-SiC composite gradient films. *Langmuir* **2018**, *34*, 1419–1428. [CrossRef]
29. Lipson, H.; Kurman, M. *Fabricated: The New World of 3D Printing*; John Wiley and Sons Inc.: Hoboken, NJ, USA, 2013; pp. 263–281.
30. Li, X.; Zhang, J.M.; Yi, X.; Huang, Z.; Duan, H. Multimaterial microfluidic 3D printing of textured composites with liquid inclusions. *Adv. Sci.* **2019**, *6*, 1800730. [CrossRef]
31. Cao, Q.; Tersoff, J.; Farmer, D.B.; Zhu, Y.; Han, S.J. Carbon nanotube transistors scaled to a 40-nanometer footprint. *Science* **2017**, *356*, 1369–1372. [CrossRef]
32. Arazoe, H.; Miyajima, D.; Akaike, K.; Araoka, F.; Sato, E.; Hikima, T.; Kawamoto, M.; Aida, T. An autonomous actuator driven by fluctuations in ambient humidity. *Nat. Mater.* **2016**, *15*, 1084–1089. [CrossRef]
33. Miri, A.K.; Nieto, D.; Iglesias, L.; Hosseinabadi, H.G.; Maharjan, S.; Ruiz-Esparza, G.U.; Khoshakhlagh, P.; Manbachi, A.; Dokmeci, M.R.; Chen, S.C.; et al. Microfluidics-enabled multimaterial maskless stereolithographic bioprinting. *Adv. Mater.* **2018**, *30*, 1800242. [CrossRef]

MDPI

St. Alban-Anlage 66

4052 Basel

Switzerland

Tel. +41 61 683 77 34

Fax +41 61 302 89 18

www.mdpi.com

Nanomaterials Editorial Office

E-mail: nanomaterials@mdpi.com

www.mdpi.com/journal/nanomaterials